Power, Culture, Economy

Indigenous Australians and Mining

Jon Altman and David Martin (Editors)

THE AUSTRALIAN NATIONAL UNIVERSITY

E PRESS

Centre for Aboriginal Economic Policy Research
College of Arts and Social Sciences
The Australian National University, Canberra

Research Monograph No. 30
2009

ANU

E PRESS

Published by ANU E Press
The Australian National University
Canberra ACT 0200, Australia
Email: anuepress@anu.edu.au
This title is also available online at: http://epress.anu.edu.au/c30_citation.html

National Library of Australia
Cataloguing-in-Publication entry

Title:	Power, culture, economy : indigenous Australians and mining / editors Jon Altman, David Martin.
ISBN:	9781921536861 (pbk.) 9781921536878 (pdf)
Series:	Research monograph (Australian National University. Centre for Aboriginal Economic Policy Research) ; no. 30.
Notes:	Bibliography.
Subjects:	Aboriginal Australians--Economic conditions.
	Aboriginal Australians--Social conditions.
	Mineral industries--Australia.
	Mineral industries--Environmental aspects--Australia.
	Mineral industries--Social aspects--Australia.

Other Authors/Contributors:
Altman, Jon C., 1954-
Martin, D. F. (David F.)
Australian National University. Centre for Aboriginal Economic
Policy Research.

Dewey Number: 305.89915

Cover design by ANU E Press.

Contents

List of Figures   v

List of Tables   vii

Foreword   ix
John Nieuwenhuysen AM

Notes on contributors   xi

Abbreviations and acronyms   xv

Acknowledgements   xvii

1. Contestations over development   1
   Jon Altman

2. Indigenous communities, miners and the state in Australia   17
   Jon Altman

3. Data mining: Indigenous Peoples, applied demography and the resource extraction industry   51
   John Taylor

4. Aboriginal organisations and development: The structural context   73
   Robert Levitus

5. The governance of agreements between Aboriginal people and resource developers: Principles for sustainability   99
   David F. Martin

6. Corporate responsibility and social sustainability: Is there any connection?   127
   Katherine Trebeck

7. Indigenous entrepreneurialism and mining land use agreements   149
   Sarah Holcombe

8. Mining agreements, development, aspirations, and livelihoods   171
   Benedict Scambary

References   203

Key Project Publications   241

List of Figures

2.1 The location of the three case studies   34

2.2 The hybrid economy framework   45

3.1 The potential recognition space for indicators of Indigenous well-being 60

3.2 Indigenous population projections 1996–2016 in selected   67
     mining regions

5.1 Gulf Communities Agreement structure   105

5.2 Key agreement governance arenas   116

7.1 The Pilbara region, Western Australia   150

List of Tables

2.1 Social indicator outcomes at mine site and some other jurisdictions,   32
     2001 Census

2.2 Social indicator outcomes in eight mining regions, 2001 Census   33

3.1 Data items available for the Thamarrurr region from Commonwealth,   56
     Northern Territory and local government agencies

3.2 Extra Indigenous jobs required in the West Kimberley by 2021   68

3.3 Summary indicative proxies of Indigenous labour force exclusion,   69
     Pilbara region, 2006

Foreword

Twenty-five years ago, in 1984, the then Chairman of the Aboriginal Development Commission, Charles Perkins, referred to the tensions between mining interests and Aboriginal opinion. These, he said, '... date back to those times of notoriety, not so long ago, when certain Aboriginal groups resisting European pressures on their land were simply swept aside ... The deep and degrading cultural disruption, the assault of noise, dust and lost privacy, the loss of social integrity of Aboriginal groups, and the outrageously low return in the way of royalties, employment and other benefits, have all formed part of the picture of the Australian development "frontier"' (Cousins and Nieuwenhuysen 1984: xii–xiii).

Despite the great rise in mining's share of Gross Domestic Product from 1.7 per cent in 1960–61 to 6.1 per cent in 1980–81, a study in 1984 found that Aboriginals then played only a small part in the operations of major mining companies; occupied mainly unskilled or semi skilled blue collar jobs; and had available to them only limited training opportunities, almost all of which were restricted to narrow job classifications (Cousins and Nieuwenhuysen 1984: 12–13).

What has happened in the quarter of a century since those findings were made? In *Power, Culture, Economy: Indigenous Australians and Mining*, edited by Jon Altman and David Martin, an excellent set of chapters enriches and broadens our understanding of this issue. A specially pleasing aspect of the project has been that PhD scholars have been included, and guided not only to completion of their degrees but also to the attainment of a prestigious publication.

It is however disappointing to learn that, after yet another major mineral boom in Australia, when in the five years to 2006 mining export revenues rose by over $100 billion (or around 70 per cent), Indigenous people still do not share equitably in the vast incomes which are generated from their lands in the remote regions of Australia. The words of Minister Jenny Macklin in 2008 that the potential of '... millions of dollars to be harnessed for economic and social advancement of native title holders, claimants and their communities ...' remained to be realised are also a sorry reflection on events in the last twenty-five years. In this comment, Minister Macklin was echoing former Minister Amanda Vanstone, who asked why land rich Indigenous people were 'dirt poor', and why the traditional owners of the land were the most disadvantaged living upon it.

If way to the better there is, it exacts a good look at the worst—after Thomas Hardy—and the authors provide a sobering set of analyses of difficulties and uncertainties in the path ahead. They also ponder the classic question of the development economics literature: does economic growth in poorer societies automatically require a new social and cultural order, or can transformation to prosperity co-exist with at least some traditional ways of living?

The answers to the questions are intractable, and a theme of the volume is the contestation of ideas and possible policy paths. Whereas a simple solution or proposal, if accepted, is presumably easier for communities and governments to act on, the scholarly review provided in this book is essential in canvassing the various options. This is the major contribution of the volume, coupled with the immense amount of new information and the broadening of the topics rightly considered relevant to the outcome.

It is a special pleasure to welcome the publication of this book, since plans for it were hatched by my former colleague at the University of Melbourne, Professor Jon Altman, in association with Rio Tinto, through the redoubtable Bruce Harvey, and the Committee for Economic Development of Australia (CEDA) where I was employed at the time. With CEDA's enthusiastic support, Jon and Bruce succeeded in arranging a successful combined Linkage Grant application to the Australian Research Council (ARC), which was coupled with substantial, generous matching funding from Rio Tinto.

There can be no doubt that the ARC Linkage Grant system, which provides incentives for industry to back up research funds from the Commonwealth, is an excellent vehicle for furthering essential enquiry effort and output. This project, under the leadership of Professor Jon Altman and the prodigiously productive Centre for Aboriginal Economic Policy Research at The Australian National University, is a prime example of the importance of the ARC Linkage Grant scheme. It is also a very favourable reflection on the enlightened attitude to funding of research by Rio Tinto, which showed absolute respect for the independence of the scholars in designing and undertaking their work and in reaching conclusions.

I therefore warmly congratulate the editors and authors on their work. They have provided material and analysis which is essential to consider in improving the longstanding unsatisfactory relationship between mining activity and the plight of the Indigenous people from whose lands the mineral wealth is being extracted.

Professor John Nieuwenhuysen AM
Director, Monash Institute for the Study of Global Movements

Notes on contributors

Jon Altman

Jon Altman has a disciplinary background in economics and anthropology. He has been involved in research in relation to Indigenous Australians since the late 1970s. In 1990 he became the inaugural director of the Centre for Aboriginal Economic Policy Research at The Australian National University, where he is still located. Professor Altman divides his research effort between a focus on national economic and policy issues and a specific regional focus on western Arnhem Land where he has worked with communities for 30 years. Professor Altman was Chief Investigator on the Indigenous Community Organisations and Miners: Partnering Sustainable Regional Development? ARC Linkage project. He is currently an ARC Australian Professorial Fellow.

Sarah Holcombe

Sarah Holcombe worked at the Centre for Aboriginal Economic Policy Research at The Australian National University for six years, to early 2008. Dr Holcombe was a part-time Research Associate on the ARC Linkage project while also working on the 'Indigenous Community Governance' project. She was then Social Science Coordinator for the Desert Knowledge CRC for two years. Before coming to CAEPR her previous research focused on applied anthropology in the Northern Territory. Her PhD research in social anthropology was undertaken in the Central Australian Luritja community of Mt Liebig (Amunturrngu), on the processes by which this settlement evolved into an Indigenous 'community'. Dr Holcombe is currently a Research Fellow at the National Centre for Indigenous Studies at The Australian National University.

Robert Levitus

Robert Levitus is an anthropologist whose primary research field since 1981 has been Kakadu National Park and surrounding areas, where he has researched social history, human ecology and land management, and applied issues around traditional ownership and Park management, sacred site documentation and protection, and the social impact of mining. He worked at the Centre for Aboriginal Economic Policy Research at The Australian National University from 1999 to 2005 on land rights reform, CDEP, and the relationship between Aboriginal development and self-determination, and completed his doctoral thesis on the sacred site aspects of the Coronation Hill dispute. Since 2005 he has worked as a private consultant on traditional ownership issues in the east Kimberley and south-west Queensland.

David Martin

David Martin is an anthropologist who worked at the Centre for Aboriginal Economic Policy Research at The Australian National University from 1995 to 2006. His research interests had a central focus on incorporating the recognition of the realities of Aboriginal cultural difference in establishing sustainable futures. They included welfare reform, community development, native title, and governance. He was the Senior Research Associate on the Indigenous Community Organisations and Miners: Partnering Sustainable Regional Development? project for its duration, working on issues of diversity and social sustainability, as well as Indigenous organisation capacity, governance and design. As a consultant, he played a significant role in establishing the organisations required to implement the Century Mine agreement.

Benedict Scambary

Benedict Scambary is an anthropologist with over 10 years experience working with the Native Title Act in the Northern Territory. In 2007 he completed a PhD at the Centre for Aboriginal Economic Policy Research at The Australian National University under the Indigenous Community Organisations and Miners: Partnering Sustainable Regional Development? project, titled *My Country, Mine Country: Indigenous People, Mining and Development Contestation in Remote Australia*. In 2005 he published CAEPR Monograph No. 25 with Dr John Taylor, *Indigenous People and the Pilbara Mining Boom: A Baseline for Regional Participation*. Dr Scambary is now the Chief Executive Officer of the *Aboriginal Areas Protection Authority in the Northern Territory*.

Katherine Trebeck

Katherine Trebeck has a disciplinary background in political science. Dr Trebeck completed a PhD at the Centre for Aboriginal Economic Policy Research at The Australian National University under the Indigenous Community Organisations and Miners: Partnering Sustainable Regional Development? project in 2005; her PhD was titled *Democratisation through Corporate Social Responsibility? The Case of Miners and Indigenous Australians*. This innovative research used a 'top down' from the company perspective versus 'bottom up' from community perspectives. Dr Trebeck is currently Research and Policy Executive at the Wise Group, a social enterprise based in Glasgow.

John Taylor

John Taylor is Senior Fellow and Deputy Director of the Centre for Aboriginal Economic Policy Research at The Australian National University. With a disciplinary background in geography and population studies, he has been researching issues related to demography, socioeconomic status, and regional profiles since 1986. He is co-editor of *Population Mobility and Indigenous Peoples in Australasia and North America* (Routledge, 2004) and the author of numerous papers concerned with Indigenous social and economic policy development. Dr Taylor was an Associate Investigator on the Indigenous Community Organisations and Miners: Partnering Sustainable Regional Development? ARC project and undertook a series of regional profiles from the perspective of applied demography in the East Kimberley, Pilbara and the West Kimberley. These were sponsored by Rio Tinto.

Abbreviations and Acronyms

ABS	Australian Bureau of Statistics
AGPS	Australian Government Publishing Service
AIATSIS	Australian Institute of Aboriginal and Torres Strait Islander Studies
ALRA	*Aboriginal Land Rights (Northern Territory) Act 1976* (Cwlth)
ANU	The Australian National University
ARC	Australian Research Council
ATSIC	Aboriginal and Torres Strait Islander Commission
ATSISJC	Aboriginal and Torres Strait Islander Social Justice Commissioner
CAEPR	Centre for Aboriginal Economic Policy Research
CDEP	Community Development Employment Projects/Program
CEDA	Committee for Economic Development of Australia
CEO	Chief Executive Officer
CHINS	Community Housing and Infrastructure Needs Survey
CRA	Conzinc Rio Tinto Australia
CZL	Century Zinc Limited
ERA	Energy Resources of Australia
GAC	Gumala Aboriginal Corporation
GADC	Gulf Aboriginal Development Company Ltd
GCA	Gulf Communities Agreement
GEPL	Gumala Enterprises Pty Ltd
GIPL	Gumala Investments Pty Ltd
GCA	Gulf Communities Agreement
GDP	Gross Domestic Product
GNP	Gross National Product
IBN	Innawonga, Banyjima and Niapali
ISS	Indigenous Support Services
LUA	Land Use Agreement
Mabo No. 2	*Mabo v Queensland (No. 2)* [1992]
MAP	Multi Level Assessment Program
MCA	Minerals Council of Australia
MDC	Major Diagnostic Code
NTA	*Native Title Act 1993* (Cwlth)
NTRB	native title representative body
PNTS	Pilbara Native Title Service
QML	Queensland Mines Ltd

RTIO	Rio Tinto Iron Ore
UNESCO	United Nations Educational, Scientific and Cultural Organisation
Wik	*Wik Peoples v State of Queensland* [2004]
YLUA	Yandicoogina Land Use Agreement
YMBBMAC	Yamatji Marlpa Barna Baba Maaja Aboriginal Corporation

Acknowledgements

The research reported in this monograph was conducted with financial support from the Australian Research Council (ARC Linkage Project No. LP0211898) and from RioTinto and the Committee for the Economic Development of Australia (CEDA) as our Industry Partners. The project's original title was 'Indigenous community organisations and miners: Partnering sustainable regional development?' and it is the question mark at the title's end that animated this project from the outset.

The project was CAEPR's first ARC Linkage Project and was undertaken by a small team of researchers between 2002 and 2007. It began in mid 2002 when David Martin, a Research Fellow at CAEPR, was recruited as part time Senior Research Associate and we recruited Sarah Holcombe an anthropologist from the Northern Land Council as Research Associate. In late 2002 we were joined by two doctoral students: Katherine Trebeck, a graduate from the University of Melbourne, and Benedict Scambary, senior anthropologist also from the Northern Land Council. These four people formed the core of the project team that undertook much of the research reported here. John Taylor was a Associate Investigator and Robert Levitus was also a part-time Research Associate for a short time after completing his PhD 'Sacredness and Consultation: An Interpretation of the Coronation Hill Dispute' at CAEPR in 2003. We thank this team for the terrific collegiality we all enjoyed while the project was under way especially evident at very crisp workshops we held with Rio Tinto and during the process of completing their chapters for this monograph.

Completing this monograph has taken a little longer than initially envisaged in part because elements of the project were completed at different times since late 2005 when Katherine Trebeck submitted her PhD thesis; and partly because members of the team dispersed to a diversity of jobs and locations. To some extent this has not been a significant issue because a number of other publications emanating from the research have been in the public domain for some time (see Key Publications at monograph end). Interestingly, the completion of this monograph is occurring after the global financial crisis of late 2008 and after the mining boom that was at its peak during the years of this project. We do not believe that the economic downturn has significantly affected any of the research findings reported here except perhaps to counsel caution to those who might see Indigenous engagement in mining as the solution to underdevelopment in remote communities.

Our research project focused on three sites and the team owes a considerable debt of gratitude to the Aboriginal people who gave so generously of their time and opinions, whether as mining company employees, agreement beneficiaries, or members of the communities in the mine hinterlands. Bruce Harvey from Rio

Tinto and John Nieuwenhuysen from CEDA who were our Partner Investigators brought considerable experience and insight to the project provided during annual workshops that were generously sponsored by Rio Tinto and through supervision and advice to doctoral students. We would like to thank them both for their ongoing encouragement and support of the project and to thank John for agreeing to write the foreword to this monograph. Personnel associated with the mining companies at each of the project locations—the then Century Zinc Limited at the Century Mine site, Hamersley Iron in the Pilbara, Energy Resources Australia at the Ranger Uranium Mine, and Rio Tinto at the Jabiluka prospect—provided invaluable information and assistance. Special mention should be made of the help given by staff of Hamersley's Aboriginal Training and Liaison Unit in Dampier.

Special thanks to Hilary Bek who provided editorial assistance and managed the monograph production process; to two external referees who provided very useful feedback; to Gillian Cosgrove for maps and proofreading; to Hannah Bulloch and Kirrily Jordan for comments; to Katherine May for proofreading; and to the team at ANU E Press who expertly produced the monograph with their characteristic efficiency.

Jon Altman
David Martin
CAEPR
June 2009

1. Contestations over development

Jon Altman

Australia is a rich first world nation. In 2007–08 it had a $1 trillion economy as measured by nominal Gross Domestic Product (GDP) with per capita income of over $A50 000. In recent years mining sector revenue has constituted a growing share of the national economy, reaching about 11 per cent of GDP in 2007–08 with a value of $A119 billion (Reserve Bank of Australia 2009). The Minerals Council of Australia (2007) estimated that in 2007–08 the value of mineral exports would reach over $A90 billion and constitute 40 per cent of Australia's commodity exports. Before the global financial crisis of late 2008, the Australian mining sector was in boom driven in large measure by the rapid industrialisation of China and India: employment in the sector had reached over 80 000 jobs by 2005–06 and was growing. Royalties paid to Commonwealth and State governments totalled $A7 billion in 2007–08; gross operating surplus in that year was $A63 billion; and in 2005–06 net profit return on average shareholder funds was 24 per cent. While some of this economic gloss may have declined in the last year, the overall significance of the mining sector to the world's fifteenth largest economy is likely to continue in the immediate future—it is as yet unclear if the current global recession is just a cyclical downturn of the business cycle or a more fundamental and structural change to economic liberalism and globalisation.

While this mining sector boom has been occurring, the share of the Australian continent owned by Indigenous Australians, what is termed the Indigenous estate, has grown to over 20 per cent of the continent (Altman, Buchanan and Larsen 2007). The areas 'owned' in various ways under land rights and native title laws are generally extremely remote and have low commercial value, except for mineral extraction. Restitution of land has come on the state's terms and excludes ownership of minerals. These lands are occupied by an estimated 20 per cent of the Indigenous population that was estimated to total just over 500 000 people in the 2006 Census.

Indigenous people, who constitute 2.5 per cent of the population, do not share equitably in the wealth of the mining sector, much of which is generated from their land in remote regions. The key issue raised in this monograph is why not?

This is a question that has also recently taxed the Minister for Indigenous Affairs the Hon. Jenny Macklin (2008) when she delivered the 2008 annual lecture to commemorate the 1992 Mabo No. 2 Australian High Court native title judgment titled 'Beyond Mabo: Native title and closing the gap'. Using language reminiscent of an earlier federal minister Amanda Vanstone she referred to 'the great

Australian paradox'. Vanstone (2007) has asked why lands-rich Aborigines were 'dirt poor' and pondered why traditional owners of land are the poorest people living on it.

Such terminology has been used in concerns raised elsewhere both internationally and with reference to Australia. For example, Karl (1997) examined 'the paradox of plenty' that haunts many mineral dependent states where production of enormous wealth coexists with extraordinary economic inequality. The paradox of plenty has similarity to the resource curse thesis (Auty 1993) that similarly queries the form of political economy (Ross 1999) that allows the existence of poverty in the midst of plenty especially during mining booms (see Langton and Mazel 2008). There is a significant literature that has explored the issue of mining and Indigenous peoples in Australia (see Cousins and Nieuwenhuysen 1984; Howitt, Connell and Hirsch 1996; Weiner and Glaskin 2007 among others, as well as Ballard and Banks 2003 for a broad anthropological sweep of the literature).

Like politicians before her, Macklin (2008) notes that 'native title is critical to economic development' and that 'properly structured property rights to land are a key component in expanding commercial and economic opportunity'. Her comments indicate she may be overlooking the issue of resource ownership under Australian law. She notes 'the potential for millions of dollars to be harnessed for economic and social advancement of native title holders, claimants and their communities', and states, 'we must not allow this potential to go unrealised'. Minister Macklin concludes: 'native title is a right which must be used as a tool to bring about positive change for social, cultural, economic purposes ... it must be part of our armoury to close the gap between Indigenous and non-Indigenous Australians'. More recently, in late November 2008, her views have been echoed by a Native Title Payments Working Group (2008) with membership drawn from a diversity of interest groups. The Working Group was convened by the Australian Government to recommend ways to ensure that resource agreements optimise financial and non-financial benefits to ensure wealth creation for traditional owners of land and the wider Indigenous community for both this and future generations.

The Minister's concerns—shared by many—do not directly tackle the inability of land rights and native title, and mining activity on these lands over the last 30 years, to rapidly reduce the socioeconomic gap between Indigenous and non-Indigenous Australians, although there has been steady improvement in most socioeconomic outcomes (Altman, Biddle and Hunter 2008).

In August 2008 mining magnate Andrew 'Twiggy' Forrest proposed the Australian Employment Covenant, a plan to provide opportunity for 50 000 Indigenous people to obtain full-time permanent work (not limited to the mining sector) within two years. The Australian Prime Minister threw his support behind the plan committing to fund the pre-employment training for all potential workers; Aboriginal rights campaigner Noel Pearson referred to the plan as 'a revolutionary breakthrough'; while Warren Mundine lauded the fact that mining companies have already come on board ('Forrest Plan to create 50 000 jobs', L. Shanahan, *The Age*, 4 August 2008). Commentators not only failed to observe that there are currently less than 50 000 Indigenous Australians in this employment category but also that only 2500 are employed Australia-wide in mining at present (Brereton and Parmenter 2008).

This monograph presents key findings from an Australian Research Council (ARC) Linkage project 'Indigenous Community Organisations and Miners: Partnering Sustainable Regional Development?' that was undertaken between 2002 and 2007 at the Centre for Aboriginal Economic Policy Research at The Australian National University. Industry partners for this research were Rio Tinto and the Committee for Economic Development of Australia (CEDA). The research was undertaken over a number of years and at a number of sites.

The key question the research sets out to address is whether major long-life extractive mines located on Aboriginal-owned land and near Aboriginal communities have the capacity to fundamentally alter the marginal socioeconomic status of Indigenous Australians in a sustainable manner. This is a question with much history both in Australia and internationally. There is considerable empirical evidence that Indigenous people rarely benefit equitably when major extractive activities occur on their customary land—indeed it is far more common for such activities to impact negatively on the livelihoods and cultures of Indigenous communities (Sawyer and Gomez 2008). Certainly in Australia there is considerable historical evidence that Aboriginal people residing in mine hinterlands in remote Australia have been socioeconomically disadvantaged (Cousins and Nieuwenhuysen 1984). At the heart of this volume is a significant policy and discourse contest between those who support the policy goal of addressing socioeconomic inequality via enhanced engagements between Indigenous Australians and miners, mediated by the state; and others who see inherent value in a livelihood approach that might be incompatible with mining, where Indigenous aspirations to live fundamentally different lifestyles play an ongoing role.

Central to this research is a growing national awareness of the dysfunction associated with inactivity and welfare dependence in remote Aboriginal communities (Pearson 2000a), and statistical evidence that the populations of remote Indigenous communities are growing quickly rather than declining (Taylor 2003). There is also growing awareness that the raft of national changes associated with 'globalisation' or 'economic liberalisation' has impacted adversely on regional Australia while simultaneously facilitating the role of corporations in providing government-like services (Gray and Lawrence 2001).

Associated with the restructuring of Australian capitalism in the late twentieth century, there has been some retreat of the state in terms of public investments in remote regions and a growing state view that profitable mining corporations have a responsibility to provide social services to remote communities, including Indigenous communities. Similarly, there has been a view expressed that Indigenous beneficiaries from agreements with mining companies should commit payments provided as compensation or benefit sharing to community purposes (Macklin 2008).

Also in the late twentieth century there were two waves of optimism that mining might provide a significant plank for what Li (2007) terms the state's 'project of improvement' in the cases examined here for remote-living Aboriginal people. The first wave was linked to the passage of land rights legislation especially the *Aboriginal Land Rights (Northern Territory) Act 1976* (Cwlth) (ALRA), and the establishment of institutions to allow payment of mining monies both to people in areas affected by mining and, more broadly, to the Northern Territory Aboriginal community. It was anticipated that the leverage provided by free prior informed consent provisions, that constitute a de facto and weak form of property right in minerals because they can be traded for consent, might result in both beneficial agreements and beneficial development outcomes (Altman 1983; Industry Commission 1991). The second wave was linked to the passage of the *Native Title Act 1993* (Cwlth) (NTA) and anticipation that right to negotiate provisions in this law would similarly facilitate negotiation for development even though de facto rights in minerals were much weaker under the NTA (McKenna 1995).

A core question raised in this monograph is whether mining has delivered improvement as a consequence of three major agreements, spread across Western Australia, the Northern Territory and Queensland, which are at the empirical heart of our investigations. Without pre-empting our more detailed findings presented in later chapters, the answer to this question seems to be that outcomes—at least as measured by standard social indicators from the five-yearly census—are disappointing. Dependence on the state remains high and indicators have improved marginally at best. This of course all depends on how one measures development. At present there is a hegemonic view of development favoured by the Australian state that narrowly construes development within the domain of the market: in accord with the strictures of neo-liberalism, development is measured with a focus on individual employment and wealth accumulation (Harvey 2007). This view is favoured by the Australian state with its focus on closing the gap in a variety of indicators including formal employment. There is an alternate view of development (Ferguson 1994; Li 2007; Scott 1998; among many others) that focuses more broadly on improvement in subjects' lives and livelihoods, but this view is currently subordinated in

Australian public discourse and policy narrative and receives limited attention. This alternate view receives considerable attention in this volume.

Part of the reason for disappointing development outcomes, whether defined in terms of social indicators or livelihoods, is linked to the limited capacity of many Indigenous organisations and communities to respond to the social impacts of mining while at the same time taking advantage of employment and other commercial opportunities. This, as we shall see, is linked in part to the ambivalence that many Aboriginal traditional owners have to mining on their land. In the three cases that we examine, land rights and/or native title laws have provided limited opportunity to actually control mining on Aboriginal lands and its social impacts on Aboriginal people. Instead, the leverage provided by new laws can best be exercised by opposing mining initially, something that is structurally facilitated by the right to negotiate provisions of the relatively new native title statutory regime. Whether opposition is fundamental or strategic (or in some cases both), when beneficial agreements are signed and mining goes ahead, relations of conflict suddenly need to transform to relations of cooperation. This switch can be difficult to manage on a community-wide or regional scale if agreements are to be beneficial (Altman 2001a).

Another reason for disappointing development outcomes is that mining companies and the Australian state (governments and bureaucracies) seem to have limited capacity to recognise the deeply-entrenched levels of disadvantage experienced by communities adjacent to remote mines and the strain on their social fabric created historically by settler or state colonisation and now, potentially, by mining. Both parties clearly see mine site engagement as providing opportunities for the state project of improvement of Indigenous people, but it is only when concerted action is taken by both parties in unison and in collaboration with Indigenous interests that positive outcomes are generated.

One of the central concerns of chapters in this monograph is the complex triangulated relationship between Indigenous people, mining companies and the state in the Australian context, an issue that has attracted some attention internationally (see Rumsey and Weiner 2004; Sawyer and Gomez 2008). Given the growing interdependence of the state and the mining sector—the former increasingly dependent on the mining sector for revenue, the latter needing licence to operate from the state—to what extent does the state operate as a neutral arbiter in its dealings with Indigenous people? This is of particular concern in situations where Indigenous people oppose mining, and the state operates as a 'broker state' (Urteage-Crovetto 2008). Another key issue raised is the proper division of responsibility for funding the services needs of Aboriginal communities adjacent to major mines. As noted above, the particular logic of the Australian state is resulting in a reduction of public expenditures on the disadvantaged living in regional and remote areas and cost shifting onto multinational corporations. While there is a growing recognition that good corporate citizenship requires major mining companies to act as catalysts for sustainable regional development, especially in the absence of other commercial options, one can also ask how far corporate social responsibility can and should stretch. Has economic liberalisation created new forms of competition for global shareholder capital that will only allow profit sharing to the minimum required to secure a licence to operate? Assuming that some share of profits can be secured for Indigenous groups, a central concern of several chapters in this monograph is what form such benefit sharing should take, and what are the mechanisms by which such benefits can be sustainably managed.

These are the sorts of questions that have animated the research reported here. This is a complex field of inquiry replete with paradoxes, like the 'paradox of plenty' referred to above and the triangulated political economy whereby the state allies with the most, rather than the least, powerful. Ultimately, this monograph seeks to address the paradox of disappointing sustainable improvement. Using the terminology of Ferguson (1994) and Li (2007) it appears that the quest for Indigenous improvement linked to mining has been 'rendered technical', as all parties have become bound up in legal agreements that are themselves frequently deficient (O'Faircheallaigh 2006). All too often as will be demonstrated, complicated social, cultural and political considerations that influence improvement have been rendered invisible. So we set out to ask what have mining companies, the state, and Indigenous regional organisations done to enhance community capacity to cope with the impacts of mining. What focus has there been on building institutional capacity to facilitate sustainable regional development and to ameliorate regional political division and conflict that is inevitably associated with mining? And what learnings have been adopted from evidence of key factors that might explain relative success or failure?

Theoretical framing

Chapters in this monograph take a variety of approaches that seek to integrate theory with detailed ethnography and local histories from the three case study sites associated with the Ranger Uranium Mine in the Northern Territory, the Yandicoogina Mine in Western Australia and the Century Mine in north Queensland.

In my view, the overarching framework of this volume is heavily influenced by the Foulcauldian concept of 'governmentality' (Foucault 1991) although individual chapters do not explicitly engage with this high level theory. Governmentality is concerned with rationalities of government or as Dean (1999: 209) puts it 'how we think about governing others and ourselves in a wide variety of contexts'. The term government is used here in a broad sense which Foucault revived from the sixteenth century to denote 'any more or less calculated or rational activity ... that seeks to shape our conduct' (Dean

1999: 209). It highlights the fact that modern political power is not simply exercised by the state, even though the state looms very large in the lives of Indigenous Australians. As will become apparent in later chapters when different authors examine the empirical messiness of the three cases from a variety of perspectives, there is a network of other actors, organisations, and enterprises that seek to guide the behaviour and decision-making of individuals. As Sawyer and Gomez (2008: 5) observe there are multiple movements of governance, both state and non-state, that aspire to fashion the conduct of people, both individually and collectively, in connection to resource extraction using various strategies, tactics and authorities. However, because the state looms so large in the lives of Indigenous Australians it plays a central role in the project to constitute Aboriginal people in the vicinity of mines as what Lawrence (2005: 40) terms 'neoliberal and job ready subjects'. It is perhaps not surprising that neoliberal governmentality (see Dean 2002) seeks to use opportunities provided by mining to convert Indigenous subjects to autonomous, responsible, employed and entrepreneurial individuals, sometimes in a paternalistic, even coercive manner (Lawrence 2005: 42).

My analysis here is framed with reference to three Australian concepts that, in my view, provide some theoretical background to the chapters in this volume. The first is Merlan's (1998) notion of interculturality, recently updated in Hinkson and Smith's (2005) edited volume; the second is Altman's (2005b) notion of economic hybridity that is evident in the articulations (and emerging tensions) between capitalist and non-capitalist forms of production in situations where a state sector looms large; and the third is Wolfe's (1999) notion of 'repressive authenticity' and Povinelli's (2002) term 'cunning of recognition' that I have found especially useful to understand the historical processes that have marginalised Indigenous subjects.

I say little about interculturality and economic hybridity here. Readers of this monograph would most likely have little difficulty accepting the view that contemporary Aboriginal social norms, even in the remotest parts of Australia, comprise a mix of customary and western (global) social norms and values. In recent years, cultural analysis in Australia has increasingly rejected false essentialised distinctions between modernity and tradition. Instead, there is a recognition of the intercultural circumstances of contemporary Indigenous life everywhere (Hinkson and Smith 2005). Clearly the precise nature of this interculturality varies enormously across the continent. It has, however, been directly influenced by the need for the codification of custom required by land rights and native title laws. Importantly, because of interculturality the nature of relations between miners and Indigenous people cannot be predetermined—in some situations mining will be embraced, in others rejected, and in most cases there is a mixed and highly contested response.

A corresponding framework for understanding contemporary Indigenous economic relations in remote regions is evident in a distinct form of intercultural economy that includes the customary (that is non-market) alongside the market and state sectors as a core element of many Indigenous peoples' livelihoods (Altman, Buchanan and Biddle 2006). Elsewhere, I have termed this the 'hybrid economy' (Altman 2005b) and it has similarities to the community economy model developed by Gibson-Graham (2002, 2005). Just as interculturality varies from place to place, so does the significance of the customary non-market sector—it is clearly most evident where people retain a close association with their customary lands and least significant where people live in urban centres and engage fully with the market economy. Economic hybridity is highly variable and greatly influenced by colonial history, environmental factors and commercial options. What is significant about the hybrid economy is not so much the size of relative sectors, but their interactions and the distinctiveness of Aboriginal economic modes informed by an amalgam of Indigenous and western norms that motivate diverse aspirations.

The hybrid economy framework is an evolving construct that seeks to explain the articulations between capitalist or market-based and non-capitalist or kin-based forms of production in situations where there is a state sector (Altman 2005b). What is important to note is that while it is based on a model of the economy with state, market and customary non-market components, it emphasises that empirically these three sectors are not only highly interdependent but also that none actually exists in isolation in some 'pure form' the overlaps between sectors are where significant productive activity occurs. In a diagrammatic representation of the model (Altman 2005b), economic activity in the four segments of sectoral overlap are illustrated to be more significant than within any discrete sector (see Altman, Chapter 2). The model also reflects the reality of a high level of contemporary Indigenous spatial and occupational mobility in remote Australia. Just like interculturality, this model does not presuppose any particular form of production and so can readily accommodate any of a range of possibilities including mine site employment.

The processes for gaining de jure rights to land involve an institutional codification of 'traditions and customs' for making claims over unalienated Crown land. For example, s.3 of the ALRA requires that Aborigines demonstrate that they are a local descent group with primary spiritual responsibility for land and sacred sites and are entitled to 'a right to forage over the land claimed'. And s.223 of the NTA requires claimants to demonstrate continuity of rights and interests under traditional laws acknowledged and traditional customs observed, and to demonstrate the maintenance of connection with lands and waters since colonisation. Through these requirements Indigenous Australians have become trapped in a western legal definition of authenticity to gain formal title to their ancestral lands. The onus is on them to prove their 'authenticity'.

These processes can be explicated with reference to the work of Wolfe (1999) and his notion of 'repressive authenticity'. Wolfe provides a critical interpretation of the history of settler colonialism and its role in transforming Indigenous Australians. He notes that the Australian settler-colonial formation was premised on displacing Indigenous people from the land rather than on any need to extract surplus value from their labour (Wolfe 1999: 1). In settler-colonies like Australia, the colonisers came to stay and so invasion is a structure—an ongoing process and not an event that occurred in distant 1788. Settler colonial societies are premised, Wolfe argues (1999: 3), on the elimination of native societies.

Wolfe's view is that expropriation of land continues as a foundational characteristic of the settler colonial society. Without going into detail about the historical state policy phases of confrontation, incarceration (on specially designated reserves) and assimilation, the state preference in recent decades has been to move Indigenous people into the settler society by 'privileging' them with the same opportunities, in theory, as those available to non-Indigenous people and thus eliminating the Aboriginal 'problem'. What is of particular relevance to this monograph is how following historical moments of social justice progressiveness (born of the failure of the Gove land rights case as well as broader shifts in Australian attitudes) in 1972 and of High Court activism in the Mabo No. 2 judgment of 1992, the Australian state has sought to define and then delineate Aboriginal entitlements to land and resources. In both land rights and native title laws, while there is some recognition that traditions and customs change, claimants' entitlements to land have required conformity to a set authenticity, as if relatively untouched by colonial history. The state, however, never seriously countenanced the provision of commercially valuable resources to claimants mainly because this would constitute a form of sovereignty, an unfathomable challenge to the logic of settler-colonial society and the Australian nation state.

'Repressive authenticity' has seen the legal recognition of land rights for some, but it has been predicated on a particular reading of colonial history and a conceptual false binary. Either Indigenous claimants are frozen in some pre-contact fiction as 'tribal' or 'traditional' (that qualifies them to claim land where unalienated and available) or else they are modern and hence are insufficiently different from other Australians to qualify for land rights. Such categorisation is misplaced. Indigenous Australians today live in an intercultural manner that can be described neither in terms of some essentialised traditionality nor essentialised modernity. The lack of recognition of this interculturality—that Indigenous people who are traditional owners of land increasingly abide by two sets of values and have aspirations that comprise aspects of both—is at the heart of the highly problematic relationship between the state and mining companies, on the one hand, and Indigenous traditional owners of land, on the other, that is the focus of this monograph.

There is a growing legal and anthropological literature that highlights the problems associated with constraining Indigenous tradition and Law to mainly accord with western legal requirements (see for example Kolig 2005; Pearson 2004; Strelein 2006; Weiner and Glaskin 2006, 2007). Consequently, even in situations where people can demonstrate continuity of tradition and custom and continual connection to unalienated land, the restrictive definitions of traditional owners means that there are winners and losers and much contestation sometimes involving the mobilisation of 'repressive authenticity' by Indigenous actors themselves (Altman 2008). There is clear empirical evidence that attachment to land remains of fundamental importance to many Indigenous people for livelihood and/or reasons of symbolic politics and identity, hence the almost total coverage of the available parts of the Australian continent by registered native title claims. As Wolfe (1999: 206–7) demonstrates, the Australian state has been effective in limiting the category of native title beneficiaries. This has effectively excluded the majority of the Indigenous population living in settled Australia from 'land rights'. As chapters in this monograph will demonstrate at a more local level, even when land rights has been granted or native title claims successfully lodged or determined, there are tensions between traditional owners as recognised by western law as 'winners' and other Aboriginal people without land rights as 'losers'.

A similar point from a somewhat different perspective is made by Povinelli (2002) using the term 'cunning of recognition' in her exposition of how the multicultural legacy of settler colonialism perpetuates unequal systems of power. Povinelli suggests that the colonised subjects are not required to identify with colonizers—something that is increasingly contestable today—but that they must instead identify with a difficult standard of authentic tradition. On one hand legislation caricatures culture as static, bounded, homogeneous and uncontested, and so laws end up shaping what they seek to recognise—Aboriginal rights to land (see Weiner and Gaskin 2007). On the other hand the state looks to atone for its past misdeeds by transforming itself into the judge of the form of (inter)cultural authenticity needed to get back land. As Sawyer and Gomez (2008: 15) note in this process the state is 'bestowing, decertifying and negating a land rights based identity to aborigines depending on whether they demonstrate appropriate, insufficient or excessive indigenousness'. This is Povinell's 'cunning' of liberal multiculturalism—it at once acknowledges difference while simultaneously disciplining, regulating and constraining otherness. It is not just the state that does this. Indigenous people too get caught up in self-monitoring and regulating who and what they are. Social struggles channelled through a discourse of identity-based rights and law can be highly problematic for 'the project of improvement' (Li 2007) as several chapters in this monograph demonstrate.

Conduct of research and the focus of contributions

The ARC Linkage project's initial aim was to look at six case studies where significant benefit sharing agreements have been signed for major extractive mines, but we ended up with only three cases, Yandicoogina, Ranger and Century. This demonstrates two things. First, the challenges that researchers face in successfully negotiating access for research purposes in what are often highly politicised environments. Second, it demonstrates the sensitivities that surround such issues at all levels—for Indigenous individuals and groups, mining corporations and the state. Our industry partner was Rio Tinto, a multinational corporation that has been at the forefront of efforts, at least in Australia and in recent times, to address how sustainable regional development might be delivered to Indigenous communities in its mine hinterlands. And yet even this powerful corporation was unable to facilitate our research in some locations like at the Argyle Diamond Mine in the East Kimberley or at the Comalco bauxite mine on western Cape York where agreements were respectively being renegotiated (see Doohan 2008) or recently implemented (see Crooke, Harvey and Langton 2006). We did not venture where unwelcome.

The resulting less ambitious focus on three major mines and associated mining agreements in three different State/Territory jurisdictions was fortuitous because, given the challenges inherent in multi-locale research, it allowed for more dense analysis. While the obvious proviso must be made that we are only examining three situations, these are nevertheless iconic sites of development contestation in the Australian context. In the chapters that follow a diversity of social sciences perspectives on these places is provided although not all authors focus on all mine sites. Nevertheless, what is provided here is unusual in the Australian context: a detailed analysis of different aspects of the economic and social impacts of mining at three different places. Such an approach has historically been undertaken by, for example, the Social Impact of Uranium Mining study (Australian Institute of Aboriginal Studies 1984) in relation to only one region (the Alligator Rivers region) but two mines (the Ranger and Nabarlek uranium mines) and two agreements.

The contributors to this volume were all involved as members of the ARC Linkage project team at various stages between 2002 and 2007; the chapters were completed at different times over the past two years and can be read either as individual essays or as part of a coherent volume. The unifying feature is that all chapters seek to provide ethnographically-grounded perspectives that complement and problematise theoretical perspectives, some of which have been outlined above.

Chapters 2 and 3 are overview chapters. In Chapter 2, Jon Altman provides analysis of the development situation of Indigenous people in Australia and their relations with the state and miners. The chapter teases out some of the structural tensions in this triangulated relationship before outlining development outcomes at the three mining regions assessed using official statistics, on one hand, and local perspectives, on the other. The chapter highlights how outcomes are differentially interpreted and contested and seeks to offer a possible solution to this problem by reconciling differing culturally-informed views of development, using case material. Rather than privileging one view of development over another, Altman optimistically argues that a model of development that can accommodate hybridity and diversity is likely to generate outcomes that are more acceptable to all parties in agreement making.

In Chapter 3, John Taylor outlines approaches to the collection of statistical data and conceptual issues that are essential to profiling the socioeconomic situation of Indigenous populations living in the vicinity of major resource extraction projects. Such data are essential to demarcate and document Indigenous populations that might be impacted by major mines, but also for ensuring that goals for Indigenous engagement with mine sites in employment and enterprise are based on a degree of statistical rigour. Most importantly, Taylor highlights that if the development impacts of mines are to be measured in any meaningful way over time, then a statistical base line is needed for comparative purposes. Ultimately, Taylor asks two crucial questions: What sustainable regional development outcomes, however measured, can be linked to a major mine, or what are the limits of mining as a driver of regional development? And what body of statistics might be needed to assess possibilities and achievements?

In Chapters 4 and 5, Robert Levitus and David Martin provide perspectives on the role of Aboriginal organisations as agents for development and as the managers of mining agreements. Both ask how a more favourable engagement between Indigenous community organisations and miners can be facilitated from different perspectives. In his contribution, Levitus considers the important role of Aboriginal organisations in their intended role as agents of development at the point of articulation between external agencies and the Aboriginal domain. In Chapter 5, Martin focuses on the governance of mining agreements and their ability to effectively deliver outcomes sustainably. In particular he highlights that agreements need to be governed as systems; that agreements need to be understood as transformative; and finally that agreements need to be seen as intercultural. Martin makes the crucially important point that governance capacity, both for Aboriginal and non-Aboriginal parties to an agreement, including governments, need to be strategically developed well ahead of agreement signing and implementation, as the transformation is made from adversarial negotiation to cooperation and collaboration after mine start-up.

In Chapter 6, Katherine Trebeck examines an aspect of her doctoral research that explores the relationship between corporate social responsibility and social sustainability, assuming that an important aspect of the latter is the capacity of

Indigenous communities to influence decisions that impact on them. She looks at four cases: the Century mine negotiations and 'sit in' (civil unrest); the Hammersley (now Pilbara) Iron's Marandoo dispute; Rio Tinto's adoption of corporate social responsibility; and the campaign to stop the Jabiluka uranium mine where community vigilance has resulted in corporate responsiveness. Trebeck's analysis focuses on agency, both within corporations where key individuals influence corporate responsiveness to community demands utilising the language of 'business case'; and within Indigenous organisations where key individuals provide crucial strategic leadership in sometimes protracted campaigns. Trebeck provides advice to corporations to assist in understanding the processes whereby Indigenous organisations represent their usually diverse constituencies, while concurrently arguing that levers can be mobilised at the community level to influence and even alter corporate behaviour.

In Chapter 7, Sarah Holcombe focuses on the nature and the diversity of Indigenous entrepreneurialism and the engagement of Indigenous entrepreneurs with mining company interests. Her focus is very specifically on Aboriginal organisations and trusts set up to manage the Yandicoogina Land Use Agreement in the Pilbara. Holcombe relates the lived experiences of a number of individuals using biographical case material and raises the crucially important issues of constraints embedded in agreements that might over-emphasise the interests of future generations over current generations thus limiting opportunities now; and also the tensions that invariably arise between individual access to benefits over group access, issues that have been at the heart of agreement making in the post-land rights era (Altman 1983). Her research highlights that Indigenous livelihood initiatives arising from mining agreements do not need to be limited to mine site employment and enterprise, there are many other ways that the finances provided in land use agreements can be mobilised to build regional social, cultural, and political capital and associated economic opportunities.

In Chapter 8, Benedict Scambary provides a summary of his doctoral research that focuses on Indigenous organisations, and specifically on the perspectives of the members of these organisational 'carapaces', in relation to mining and its potential for sustainable local development. Scambary is particularly interested in the impacts of beneficial mining agreements on Indigenous livelihoods in mine hinterlands and focuses on three, the Ranger Uranium Mine Agreement, the Yandicoogina Land Use Agreement and the Century Mine Agreement across three jurisdictions. He provides a critique of the efficacy of such agreements, arguing that a combination of the depth of Indigenous disadvantage and the mainstream terms of the agreements themselves hamper their potential to deliver sustainable outcomes for Indigenous people associated with all three agreements. In his chapter he argues that a fundamental limitation of these mining agreements is their incapacity to engage with and augment the diverse livelihood objectives of Indigenous people which results in ambivalent responses to mining on their part. Scambary suggests that successful engagement between the mining industry and local Indigenous people who reside in mine hinterlands is dependent on accommodation of existing Indigenous skills and knowledge. Examples abound from across all three locales he studies of Indigenous people successfully striving to engage in multifaceted ways with the mainstream economy, and the mine economy, whilst not compromising their cultural identity and aspirations.

Conclusion

As this monograph is completed, Australia is in the grips of the global financial crisis and an associated global economic recession. These events might encourage, one might imagine, a more critical thinking about the role that mining might play in ensuring sustainable Indigenous development. But there seems to be little evidence of a healthy scepticism about the risks that a development pathway closely linked to mining might entail. The Rudd Government has made Closing the Gap its key Indigenous affairs policy focus, with a halving of the employment gap in the next 10 years one of its key aims (Council of Australian Governments 2008). There is no doubt that a greater engagement of remote-living Indigenous people in mine economies is seen as a central plank of this goal that will require 100 000 new jobs in the next decade.

It is perhaps no coincidence that the Australian Employment Covenant is the brainchild of mining magnate Andrew Forrest. It is also of no surprise—but of some disappointment, as noted at the outset—that Minister Macklin (2008) focuses primarily on how Indigenous people should spend their agreement payments on community benefit without any engagement with the issue of whether such payments are adequate or equitable—arguably *the* crucial politico-economic question—or whether traditional owners of land should be required to quarantine their negotiated compensatory payments for damage to their land for wider community benefits. These are issues that have had a long history in Indigenous affairs policy debates (see for example Altman 1983; Cousins and Nieuwenhuysen 1984; O'Faircheallaigh 2004a, 2006; Trigger 2005).

Mining is fundamental to the wealth of Australia and as the Indigenous estate has expanded to cover over 20 per cent of the continent there is no doubt that more and more exploration and mining will occur on Aboriginal land. However development outcomes are defined—narrowly as conforming to a mainstream ideal or more broadly to focus on a livelihood—there is no doubt that for decades now these outcomes have been disappointing from Indigenous, corporate and state perspectives. It has been surprising just how little rigorous research is undertaken around Australia on the socioeconomic impacts of mining. While there are numerous 'top–down' statements about the benefits that mining should deliver, grounded Indigenous viewpoints are under-reported or unheard; pressure is mounting on Indigenous representative organisations to view mining as the panacea for regional and Indigenous development.

Yet even within the three cases on which chapters in this volume focus there are some clear variations in engagement with, and the impacts of, the mining sector that are heavily influenced by history, the nature of the local Indigenous polity and land ownership, and the diversity of Indigenous responses. This in turn has been greatly shaped by provisions in Australian land rights and native title laws that play a crucial role in (predictably) creating regional Indigenous diversity and political conflict by differentiating traditional owners or native title groups from others living in the region.

Chapters in this volume do not assume that mining is either positive or negative—there is no advocacy for any particular outcomes in situations where mining occurs on Aboriginal-owned land—but rather they set out to explore what has happened at three mine sites from a diversity of perspectives. It is recognised, though, that mining will be a site for contestation and that development, however defined, will only occur if Indigenous organisations are empowered and have capacity to negotiate and manage beneficial agreements in accord with local and regional Indigenous aspirations which may themselves, as we shall see, differ. These aspirations will, in all probability, be diverse so that not all Aboriginal people will seek employment opportunity at mine sites. Indeed, in some situations Indigenous people might actually seek mining employment so as to earn incomes to enhance opportunities for futures on country. In other situations even massive long-life mines with heavy Indigenous mine site engagements will generate insufficient opportunity to solve Aboriginal development problems, even if mining were a sustainable regional prospect.

An important question that is raised in this monograph from a social sciences perspective is whether there is too much collusion between the state and mining companies that offers a development pathway—concentrated mine dependence—that is too risky and too divorced from the preferences of many Indigenous subjects. At the current historical moment, perhaps the choice offered to remote living Indigenous people is too influenced by the dominant logic of neo-liberalism: engage with the mainstream as individual subjects or miss out. Such a stark choice seems to be at loggerheads with other more flexible options that are essential for sustainable livelihoods for people living on their lands and in accord with contemporary diverse, but intercultural, preferences. One such option is encapsulated in the notion of 'economic hybridity'. The simplistic choice, modern or customary, that so dominates Australian public and intellectual debates and that ignores the intercultural needs to be challenged by a combination of grounded realism and engagement with a subordinate development discourse that is all too rarely articulated in Australia today. The essays in this volume ultimately aim to broaden this development debate, while also providing some insights into how, when mining does occur, one might look to better outcomes for all parties.

2. Indigenous communities, miners and the state in Australia

Jon Altman

Introduction

Economic globalisation, as noted in the previous chapter, has served Australia well in recent years. Until 2008 this rich developed country was in the grip of an export-oriented resources boom. Even as much of the country was in severe drought, the economy continued to expand uninterrupted for 15 years. Australia, a country of only 20 million people, is the world's fifteenth largest economy. This good economic fortune has not been shared by all Australians. The Minerals Council of Australia (MCA) (2006) notes that 60 per cent of minerals operations in Australia have neighbouring Indigenous communities, but is concerned that the Australian Government is not doing enough to support the development of Indigenous representative structures or investing enough in community infrastructure and social services in remote and regional Indigenous communities. The MCA has identified a need for 70 000 additional employees by 2015 and an opportunity for an alliance between government, the private sector and Indigenous communities that will allow social, employment and business opportunities in mining regions to be taken up. While 2008–09 has seen a substantial number of job losses in the mining sector as a result of the global recession, demand for mining workers is likely to rapidly increase during the next phase of global economic growth. As noted in Chapter 1, an optimistic employment outcome is shared by the Australian Employment Covenant, although it too might be dampened somewhat in reality by the global financial crisis. Nevertheless, in recent years the MCA has shifted its rhetorical emphasis from litigating native title to building sustainable Indigenous communities.

This perspective can be contrasted with the somewhat paternalistic state approach to Indigenous engagement with the mining industry—promulgating a monolithic and mainstream development approach and trajectory. Historically, wealth creation for most Australians has been predicated on expropriation of Aboriginal lands, initially for agriculture and then also for mining from the nineteenth century. In the 1970s, new and progressive land rights laws in the Northern Territory saw much land returned to Aboriginal people, but without mineral and other resource rights. At most, Aboriginal traditional owners were accorded free prior informed consent provisions, but even these could be overridden by national interest provisions.

Since then, and despite the 1992 Mabo No. 2 Australian High Court decision and native title laws, Indigenous resource rights have been diluted: the *Native*

Title Act 1993 (Cwlth) (NTA) framework provides a statutory right to negotiate but no requirement for Indigenous consent to mine. In the last decade, and especially between 2004 and 2007, the Howard Government controlled both Houses of Parliament and moved to disempower Indigenous communities and to depoliticise their institutions. Counter to emerging international conventions, especially those in the United Nations Declaration on the Rights of Indigenous Peoples (paradoxically supported by the Rudd Government on 3 April 2009) special Indigenous rights to land, livelihoods and resources are not given much support by the Australian state.

The dominant development approach is now focused on statistical equality for Indigenous Australians. An emergent narrative has looked to pillory the past policies of 'self determination' as the cause of the contemporary Indigenous absolute and relative socioeconomic disadvantage that can be tracked with statistical social indicators from each five-yearly census since 1971 (Altman, Biddle and Hunter 2008). The new emphasis in situations where there is mining activity is to emphasise the development opportunity that this presents to Indigenous communities.

The Indigenous response to this development focus has been mixed. In some situations mainstream opportunity has been embraced, in others fiercely opposed: to some extent the two extremes have resulted from the absence of an alternate development discourse where diverse Indigenous views about development can gain a legitimate hearing. Such extremes have also been structured by the nature of statutory provisions for engagement that provide incentive to leverage benefit from adversarial opposition. The absence of support for alternative livelihood options has contributed too, with Indigenous communities often facing the choice between mining or welfare dependence—and increasingly even welfare is being withdrawn if the mining option is not taken up.

The historical legacy of neglect, poor health, poor housing and poor education will make Indigenous economic and social integration into the mainstream mining sector extremely difficult. Somewhat paradoxically, to be in a position to reclaim traditional lands, claimant groups under the NTA framework must demonstrate before the courts continuous connection to the land and the maintenance of custom. At the same time active engagement with the mainstream mining sector could jeopardise the maintenance of custom and connection bases for claim, while the latest case law indicates that mining leases extinguish native title and that the 'bundle of rights' that constitute native title do not include contemporary commercial mineral rights.[1]

---

[1] See Strelein (2006: 59–77) and her discussion of the redefining of extinguishment in the *Western Australia v Ward* High Court decision in 2002. See Glaskin (2003) on the bundle of rights approach to native title.

This chapter examines the development situation of Indigenous people in Australia, focusing on the interactions between them and mining sector multinational corporations in a developed nation state; and the mediating, regulating and limiting role fulfilled by state governmentality. In a somewhat inconsistent manner that will be investigated, the Australian state—that is committed to economic liberalism in such processes—seems to be voraciously pursuing a strategy to further disempower Indigenous people in an extremely uneven power relationship.

Initially, I will provide an analysis that focuses separately, and at a general level, on the historic and current approaches of the state, miners and Indigenous people to economic development. I will then draw more specifically but briefly on case study material from the three major mines: Yandicoogina in the Pilbara, Western Australia; Ranger (and the adjoining Jabiluka prospect) bordering Kakadu National Park in the Northern Territory); and the Century mine in the Gulf region of north Queensland. These three major resource development projects respectively mine iron ore, uranium and zinc and lead.

Whether the opportunities presented by these major mines on Aboriginal land have generated development outcomes, and according to whose criteria, is a major issue that animates the analytical section on development contestation. The notions of 'power' and 'identity' loom large in this analysis. A final section challenges the currently dominant view of development focused on 'practical reconciliation' (a term coined by the Howard Government largely to discredit the symbolic aspects of reconciliation) and Closing the Gap (that is, socioeconomic equality) by exploring an alternate view that emphasises choice, self determination and cultural continuity, possibly at the expense of material prosperity (Altman and Rowse 2005: 176).

Scene setting: The overall context

Australia is a continent of some 7.7 million square kilometres that until 1788 was occupied by between 250 000 and 750 000 Indigenous peoples. Colonisation of the continent occurred incrementally over the following 150 years, with this process clearly impacting on the ability of Indigenous peoples to maintain links with their ancestral lands and pre-colonial economic base. As land was expropriated, Indigenous marginalisation and relative poverty was created. While the converse view that return of the land will somehow magically fix the Indigenous development problem today has some intuitive appeal, it also has major shortcomings.

Australia is a liberal democratic federation of eight States and Territories with a population of 20 million people, just over 500 000 of whom are Indigenous (just over 2 per cent of the total). Reliable information on Aboriginal and Torres Strait Islanders was not readily available before 1971. It was only after an amendment to the *Australian Constitution Act 1901* in 1967 that s.127, which excluded Indigenous Australians from the census count, was repealed. From the 1971 Census, Indigenous people could self-identify and were included in this manner in official statistics. In each of eight censuses the segment of the population that can be identified as Indigenous (owing to the inclusion of a voluntary self-identifying question on Aboriginality from 1971) has fared far worse on socioeconomic indicators than the non-Indigenous population. Numerous analyses since the late 1970s (summarised in Altman and Rowse 2005; Altman, Biddle and Hunter 2008) have highlighted this, although many also caution against reliance on culture-relative 'western science' social indicators. Information from five-yearly censuses over 35 years is analysed in five broad areas—employment, education, income, housing and health—in Altman, Biddle and Hunter (2008). The data indicate that while absolute Indigenous well-being at an aggregate national level has improved, relative well-being has improved much less and has stagnated in recent years. The one outstanding indicator of Indigenous disadvantage that has barely shifted since 1971 is Indigenous life expectancy. This is estimated at 17 years lower for Indigenous Australians compared to the rest of the population, at the national level.

Australia is a so-called strong state, covering a minerals-rich continent. In recent years it has established a reputation as a country with low sovereign risk: its relative isolation that historically constituted a competitive disadvantage is now a positive in terms of isolation and freedom from risk of terrorism. Yet within this rich, post-colonial, post-industrial society is a marginalised Indigenous minority, some of whom live in conditions that are far worse than official statistics and averaging depict.

The state

The state[2] has loomed large in the lives of Indigenous Australians. Colonisation in 1788 was an instrument of British state policy. The history of state policy is greatly complicated by the emergence in the nineteenth century of an Australia compromising six colonies each of which developed its own policies for dealing with its Indigenous inhabitants. Generally, special laws set Indigenous Australians apart from other colonial citizens for their 'protection': the purpose of this can be interpreted optimistically as either a means to prepare Aborigines for future full citizenship or to 'smooth the pillow for a dying race' (Altman and Sanders 1991: 1). After federation of the six colonies into the Commonwealth of Australia in 1901, the new State governments continued to manage their

---

[2] The nature of the state is complicated in Australia by the federal system that views Indigenous people as both citizens of the Commonwealth of Australia and of States and Territories. In general, mining is regulated by State and Territory governments, but the focus here is on the Commonwealth government and its bureaucratic apparatus and institutions, in part because it sets the broad policy frameworks for mining and Indigenous affairs.

Indigenous minorities; the Australian Constitution excluded the Federal Government from an active role in Indigenous affairs, except in the Northern Territory which was ceded to the Commonwealth from South Australia in 1911.

By the 1950s it became clear that the Indigenous population was not disappearing; it was officially estimated at between 70 000 and 80 000, representing about 1 per cent of the total Australian population. From then assimilation became the central term of policy although it was only officially defined in 1961. The policy of assimilation operated by providing exemptions for individual Aborigines from the special bodies of laws that continued to be exercised over all Aborigines (until they could prove worthiness for normal citizenship). Assimilated Aborigines could take their place as full members of Australian society without any separate legal status. In some jurisdictions colonial state policy sought to centralise, sedenterise and assimilate Indigenous people in government settlements and missions on reserves up until the early 1970s. By then there was mounting evidence that the policy of economically and culturally integrating Indigenous people into the majority mainstream society was not working, especially in remote regions where state colonisation had arrived relatively late, in some cases in the 1950s.

It was only from the 1970s that federal approaches to Australia's Indigenous minority underwent significant change, with self determination becoming the central term of Indigenous affairs policy for a short period. Suddenly there was a rapid escalation in federal involvement in Indigenous affairs including a dedicated government department, an elected Indigenous representative organisation, Indigenous specific programs, the establishment of thousands of community-based organisations to locally administer programs, and a bold start in the creation of laws to enshrine land rights for Indigenous Australians.

It is the last development that is of greatest significance to the discussion here. The Woodward Land Rights Commission of 1974 (Woodward 1974a, 1974b) recommended a statutory land rights regime for the Northern Territory, a jurisdiction still administered from Canberra. This was a political commitment by the incoming Whitlam Government to address the perceived social injustice of the Gove land rights case where the plaintiffs sought to stop the development of a massive bauxite resource development project in north-east Arnhem Land. The plaintiffs lost the case in the Northern Territory Supreme Court in 1970, but just six years later the *Aboriginal Land Rights (Northern Territory) Act 1976* (Cwlth) (ALRA) was passed. This statute created a special form of land title, communal inalienable freehold title, to be held by land trusts and managed by statutory authorities called land councils. While the Woodward Land Rights Commission did not recommend conferring of legal mineral rights on land owners, it did recommend that Indigenous traditional owners of land have a right tantamount to free prior informed consent over commercial development. This is often referred to as the right of veto.

The ALRA had many innovative and important provisions including a land claims mechanism (although this is limited to unalienated Crown land) and a statutory financial framework that saw the equivalent of mining royalties raised on Aboriginal land shared between people in areas affected by mining (30 per cent), the administrative costs of land councils (40 per cent) and other Indigenous interests in the Northern Territory (up to 30 per cent). The right of veto constituted a tradable de facto property right in minerals because there is provision to negotiate payments for consent (see Altman 1983; Industry Commission 1991). The establishment of land councils provided traditional owners with statutory authorities empowered to claim land on their behalf and to represent them in negotiations with resource developers. Meeting the administrative costs from mining royalty equivalents paid by the Commonwealth was an innovative means to provide them budgetary certainty and *a degree of independence* from the government of the day.

Subsequently some other Australian States passed land rights laws, but none have been as comprehensive as the ALRA.[3] While the Whitlam Government had intended this land rights law to be the benchmark for other States, no subsequent Federal government was willing to introduce national land rights. Indeed an attempt to do so in 1984 by the Hawke Labor Government became an electorally contentious issue, especially in Western Australia where proposals for State land rights law following the Seaman Inquiry were jettisoned by the Burke State Labor Government after a virulent anti-land rights campaign mounted by the Western Australian Chamber of Mines (Libby 1989).

The 1992 Mabo judgment, that jettisoned the concept of *terra nullius* and recognised that Indigenous native title could exist in Australian common law, was the precursor to the NTA. This law was as much about validating existing non-Indigenous interests in land as about providing opportunity for native title to be claimed over unalienated Crown land and pastoral leasehold land. The State where native title was most likely to be determined was Western Australia because over 34 per cent of the State (or 863 000 square kilometres) remained vacant Crown land in 1993. Consequently, the Western Australia Government initially refused to accept Commonwealth native title law and instead passed it own Land Titles and Traditional Usage legislation that was challenged and deemed unacceptable in the High Court. Pastoral leasehold land that covers over 40 per cent of Australia was shown in the High Court Wik decision of 1996 to be potentially compatible with native title. This resulted in amendments to weaken native title law in 1998 by the Howard Government largely to privilege

---

[3] For a detailed recent summary see Aboriginal and Torres Strait Islander Social Justice Commissioner (ATSISJC) (2006).

the commercial interests of miners over native title interests in the name of workability.

Strelein (2006) has recently referred to native title case law since Mabo as 'compromised jurisprudence'. It is extremely complicated law that has nevertheless seen native title determinations (some exclusive possession, some partial) over an estimated 8 per cent of Australia (Altman, Buchanan and Larsen 2007). Native title has seen the establishment of new institutions including native title representative bodies (to represent native title claimants); a National Native Title Tribunal based in Perth, Western Australia to register native title claims (that then are heard in the Federal Court) and to mediate Indigenous Land Use Agreements; and Prescribed Bodies Corporate to hold native title in perpetuity once determined. The emergent institutional landscape is extremely complex.

What is especially significant is that the native title legal framework provides native title interests (claimants and holders) no special rights over minerals beyond negotiation rights. The so-called right to negotiate is just that but if agreement to proceed cannot be met in a strictly stipulated time frame of six months, then an arbitral process allows mining to proceed with compensation determined without recourse to the value of minerals. This mechanism is intended to hasten agreement making between resource developers and Indigenous parties, with commercial deals possible within the six-month window of opportunity with registered claimants as well as determined native title groups. To date, most agreements have been made with claimants.[4]

It is estimated that 1.5 million square kilometres is now held under some form of Indigenous title. As Fig. 2.1 shows, these land holdings are heavily skewed in favour of regional and remote Australia owing to colonial history that saw only land of low commercial value reserved for Aboriginal use or unalienated. It is estimated that only about 100 000 Indigenous people, or about 20 per cent of the 'officially enumerated' Indigenous population, live on their now legally recognised ancestral lands increasingly referred to as 'the Indigenous estate' (see Altman, Buchanan and Larsen 2007).

What is especially important for the discussion is the broad new direction that was taken by successive Howard Governments from 1996 to 2007, during which time there emerged a narrative to the effect that policies over the past 30 years have failed to deliver 'development' and that land rights and native title laws introduced for a number of broad rationales (see ATSISJC 2006: 16ff) are implicated in this failure. The ALRA is the iconic 'high water mark' statute that has been targeted for special focus with a major review (Reeves 1998) recommending fundamental changes, subsequently rejected by a Parliamentary

---

[4] See the comprehensive Agreements, Treaties and Negotiated Settlements Database accessed 22 February 2007, <http://www.atns.net.au/>.

Inquiry (House of Representatives Standing Committee on Aboriginal and Torres Strait Islander Affairs 1999).

These proposed amendments were reintroduced in 2004 when it became clear that the Federal Government would gain rare absolute power with control of both Houses of Parliament for the period July 2005–November 2007. Amendments to the ALRA passed in 2006 were aimed at facilitating enhanced exploration and mining on Aboriginal owned land as made clear in Explanatory Memoranda.[5] It was assumed that benefits would trickle through to Aboriginal communities, a view that has been challenged by the Productivity Commission (Steering Committee for the Review of Government Service Provision 2005). The policy approach here is confused and contradictory: Indigenous representative organisations have been weakened; and incentives to consent to mining have been reduced, as benefit streams need to be applied to community purposes that are arguably the domain of the state. At once neoliberal policy is highlighting the need to incentivate individuals and families, while overlooking that it is not only excluding individuals from direct benefit but also failing to recognise alternate cultural and environmental values of the Indigenous estate.

Multinational corporations

Mining in Australia is dominated by large multinational corporations, as evident by listings on the Australian Stock Exchange and membership of the principal industry lobby group the MCA[6] : it is estimated that 80–85 per cent of mineral production in Australia is undertaken by multinational corporations.

Historically in Australia mining companies have allied with the state (both Commonwealth and State/Territory) to override Indigenous views about resource development on their lands. There are some situations that have seen extraordinarily unbalanced conflicts between the pro-mining state and large mining companies on one side and the opposing Indigenous communities on the other. The Gove case, where the mining company Nabalco was joined by the Australian Government in opposing Milrrpum and others, has already been mentioned. There have been other examples, such as Comalco on western Cape York where the Aboriginal community of Mapoon was forcibly closed and its residents removed to make way for mining in 1963; Noonkanbah and the Argyle Diamond Mine in the Kimberley region; and iron ore in the Pilbara region of Western Australia. In a landmark study in the early 1980s, Cousins and Nieuwenhuysen (1984) looked at each of these sites and illustrated how Indigenous people with limited, at times non-existent, rights opposed mining

---

[5] 'The principal objectives are to improve access to Aboriginal land for development, especially mining … ' (Parliament of the Commonwealth of Australia 2006: 3).
[6] See <http://www.minerals.org.au/> and especially the MCA's Board of Directors dominated by major multinational corporations; the MCA is funded by member contributions made on a sliding scale based on economic size.

on their lands. They also illustrated with quantitative data that Indigenous peoples benefited little from major mining activity on their lands and adjacent to their communities. In all the cases of conflict, the state, keen to see resource development and associated regional development (but not necessarily for Indigenous people), sided with miners.[7]

The mining industry's relations with the state in terms of land rights and native title have waxed and waned in the last 30 years but have fundamentally been oppositional, seeing consent or negotiation provisions where the state has supported land rights as just another regulatory hurdle in the way of unimpeded access to Aboriginal land for mineral exploration and exploitation. The Australian Mining Industry Council (as the MCA was then known) strongly opposed land rights in the 1970s, strongly opposed national land rights in the early 1980s, and mounted a very effective campaign (with television advertisements) against Western Australia land rights in 1984 (see Libby 1989). Similarly the mining industry strongly opposed the passage of native title legislation in 1993 and made representation for the dilution of land rights and native title in the 1990s. This was especially evident in the lead up to amendments to the NTA in 1998 that effectively removed the right to negotiate for native title groups on the pastoral estate that covers 40 per cent of Australia.

Much of this changed from the mid-1990s with one company, Rio Tinto, taking a strong leadership role. Its then chairman, Leon Davis, challenged the mining industry to work within the existing native title statutory framework rather than continuing to oppose it (Davis 1995).[8] There has also been a growing recognition of the extent of Indigenous disadvantage and the contemporary strains on communities' social fabric associated with large scale development. The Rio Tinto booklet *The Way We Work* (published regularly) indicates a cautious recognition that mining companies might need to be the catalyst for sustainable regional development, especially in the absence of other commercial opportunity. This corporate shift has been influenced by the weak leverage that native title law provides, as well as events at Bougainville and Ok Tedi in nearby Papua New Guinea which demonstrate the complexity of issues circulating around national sovereignty, democratic governance, local community and business autonomy (Kirsch 2006).

Improved global communications and globalisation have made mining companies more accountable transnationally and have made competitive advantage partially

---

[7] The one exception to this general rule occurred with the Coronation Hill dispute, where the Hawke Government in 1991 accepted the recommendation of the Resource Assessment Commission to disallow mining in the southern part of World Heritage Kakadu National Park. This exceptional decision was based on both environmental grounds and Indigenous opposition on cultural/religious grounds. It was a *cause celebre* that resulted in a considerable political backlash.

[8] Research about Rio Tinto reported here has focused on some of its operations in remote Australia, and none of its operations in settled Australia or globally.

dependent on community relations track record. Concomitant with the increased power to trade that economic liberalism has brought to transnational capital is a growing international interest in how mining companies might demonstrate a commitment to socially sustainable development. However, the ethical requirement to profit share rarely extends beyond the requirement to gain licence to operate.

Harvey (2004) outlines how, in the Australian context, Rio Tinto is keen to operate as a good corporate citizen and in a socially responsible way. Rio Tinto's recent policy changes mean that it now seeks to make agreements with Indigenous peoples well-represented by their own organisations, with agreements always signed off by the state (usually with major agreements at both Commonwealth and State/Territory levels). Modern agreements seek to deal directly with local Indigenous land owners, although as we shall see below, this can be a problematic interest group to unambiguously define and target; and to ensure that benefits are utilised for regional development. In parts of remote Australia, it is large mining companies that sometimes make the significant social investments that are the responsibility of the state, in part because historically the development of new mining towns was a condition of licence to operate. As Holcombe (2006) notes, in regions like the Pilbara these towns were initially closed to Indigenous people, being earmarked for mine site personnel only.

In the past the peak mining lobby group, now the MCA, often sided with the state counter to Indigenous interests, especially in relation to diluting land and native title rights (as noted above). Through the MCA, some mining companies have had the ability to influence state policies that have resulted in changes to laws that have directly benefited their interests.[9] More recently, the MCA has developed an overarching position of looking for strategic partnerships with Indigenous communities and leaders, the latter through its Indigenous Leaders Dialogue, and by sponsoring meetings between the National Native Title Council (the peak association for native title representative bodies), the MCA and senior industry representatives.[10] In its Indigenous Relations Strategic Framework, the MCA notes that there are now over 300 agreements between minerals companies and Indigenous communities throughout Australia. By necessity the MCA interacts with remote Indigenous communities, many of which have poor public service delivery owing to remoteness. There are limited mainstream economic development opportunities at such communities and often the minerals industry claims that it is the only vehicle for socioeconomic development (MCA 2004).

---

[9] Also, some companies operate less ethically than others, as noted in an assessment of 45 mining agreements by O'Faircheallaigh (2004a).
[10] See <http://www.minerals.org.au/>.

The challenges that the minerals industry faces in its relations with Indigenous communities though are significant. Some are legal and structural—for example, a recent High Court decision in *Western Australia v Ward* judged that historic mining leases extinguish native title, which places mining companies and Indigenous claimants in direct political conflict. Similarly, the inevitable environmental damage caused by mining activity places it in conflict with those Indigenous interests that accord high value to the cultural and physical landscape. Also, the minerals industry defines economic development in a similar vein as the state: in western and mainstream terms of jobs, commercial opportunities and incomes. The industry is increasingly recognising that it needs to be careful not to take on state-like functions, because meeting the significant backlog could jeopardise a project's profitability. This is evident in the MCA's increased willingness to criticise governments for poor delivery of essential community social and physical infrastructure like education and health services and housing and water (MCA 2004: 5).

The new positioning of the MCA suggests that its members are more willing to ally with Indigenous communities in their quest for greater needs-based service delivery and independent representative bodies (MCA 2006). However the fundamental issue of what might constitute the sustainable economic legacy that the mining industry might leave Indigenous communities is rarely addressed. It is clear that even if all opportunities at mine sites were taken up by Indigenous people, eventually the mine will close—so mine dependence, like state dependence, can be very risky.

Indigenous rights and economic levers

As already noted, while the 2006 Census estimated the size of the Indigenous population at just over 500 000, the colonisation process and the subsequent mechanisms for returning land to Aboriginal people under land rights and native title laws have meant two things. First, returned land has by and large been limited to unalienated Crown land in an inverse relationship to the colonial settlement pattern.[11] Consequently, most is remote and has very limited commercial value. Second, returned land is very inequitably bestowed. Exact figures are not available, but in spatial terms Indigenous land holdings are only significant in three jurisdictions: the Northern Territory (just over 600 000 square kilometres), in South Australia (just over 200 000 square kilometres), and in Western Australia (just over 360 000 square kilometres). These figures are conservative and do not include native title determinations that are seeing Aboriginal reserves and leasehold in Western Australia rapidly converted to

---

[11] There is a history of programs established by the state to purchase land for Indigenous groups, mainly for commercial pastoral and agricultural purposes. In 1995, an Aboriginal and Torres Strait Islander Land Fund was established to provide land for Indigenous people whose native title had been extinguished, see <http://www.ilc.gov.au/>.

exclusive native title possession, often after lengthy litigation.[12] In New South Wales, Victoria, Tasmania and the Australian Capital Territory there is little Indigenous owned land, with some in recent times being purchased and then vested with Indigenous traditional owner groups.[13]

There are no data that accurately quantify the total number of Indigenous people who are land owners or who live on Aboriginal land. Two proxies are often used. The first is the number of Indigenous people residing at about 1200 discrete Indigenous communities mostly located on Indigenous owned or managed lands. This figure is estimated in official statistics at between 100 000 and 120 000 (Altman 2006). The second is the number of Indigenous people living in regions classified in census geography as remote and very remote; this figure is just over 120 000 (Taylor 2006: 5). It is these people who interact most frequently with mining companies in relation to exploration and mining on their lands.

As already noted, land rights and native title provide a variety of weak forms of property (resource) rights to Indigenous land owners, ranging from full property rights to a limited range of minerals in two jurisdictions (New South Wales and Tasmania) at the strong end, to a mere right of consultation under amended native title law at the weak end. In reality, the strongest property rights are the right of consent provisions of the ALRA. Even here problems remain owing to the conjunction between exploration and mining[14] and the absence of opportunity to disengage if the impacts of long-life intergenerational mines are socially or environmentally negative. Again two qualifications are important (Altman 2001a). First, all land rights and native title laws passed in Australia have guaranteed that existing mining interests will prevail; so-called 'prior (commercial) interest'. Second, a series of High Court decisions in land and sea native title claims to date have ruled that commercial (usually non-Indigenous) interests always take precedence over customary Indigenous interests.

The clear message is that Indigenous interests are subordinate to commercial interests in Australian society today. Yet the mechanisms available if Indigenous interests want to engage commercially are limited, as if the dice are doubly loaded against Indigenous people.[15] For example, a shortcoming of the available political and economic levers to consent or negotiate is not only that they can

---

[12] See native title claim and determination maps at <http://www.nntt.gov.au/>.

[13] For a spatial summary see Fig. 1, Pollack (2001) and Altman, Buchanan and Larsen (2007); for a discussion of land rights regimes see ATSISJC (2005) and for native title determinations see the National Native Title Tribunal website <http://www.nntt.gov.au/>.

[14] While in Australian policy and popular discourse it is common to speak of the Indigenous right to veto mining, in reality Indigenous interests at best have a right to veto exploration; changes to Northern Territory land rights law in 1987 meant that traditional owners could not exercise the veto over mining once exploration had been approved.

[15] For a recent analysis of the quality of agreements struck see O'Faircheallaigh (2004a) and for an analysis of the imbalance in arbitrated decisions see Corbett and O'Faircheallaigh (2006).

be overridden, but that they are most effective at the pre-production phase of a resource development project.

Interestingly the dominant neo-liberal focus on individualism, entrepreneurship and profit maximisation does not extend to Indigenous peoples in the mining context. Under all statutes there are some—at times hotly contested—legal mechanisms for land owners whose lands are affected by mining to distribute any compensatory benefits from mining companies to other regional Indigenous interests, irrespective of social or economic affectedness. The purpose for which compensation is paid is unresolved in Australian law and policy. As has been discussed elsewhere (Altman 1983, 2001a) it is unclear if compensation is paid as a share of mineral rent (in which case it should go to land owners) or as a means to offset social and economic disruption, in which case it should be targeted to such purposes.

In the case of the native title legal framework it is clear that compensation is payable for impairment of native title rights, but mechanisms embedded in the right to negotiate process actually seek to expedite agreement making by getting resource developers and native title interests to make agreements rather than seek compensation determinations by the National Native Title Tribunal or court system. In this process, the value of a mine can be considered in agreement making, but not in compensation determinations. The very fact that the language of compensation is used suggests that there is an expectation of negative impacts from resource development (Altman and Pollack 1998).

Indigenous aspirations for engagements with mining companies are highly variable, even for traditional owner groups and other Indigenous interests in mine hinterlands. It is not surprising to have a diversity of views ranging from pragmatic acquiescence to mainstream views of development, to total opposition based on fundamental and at times culturally-based opposition to mining, unavoidable environmental damage, and the possible desecration of sacred sites. It is rarely appreciated that for many Indigenous groups the landscape is both a source of livelihood and the essence of the Dreaming, the sentient landscape being created according to Indigenous ontology by Ancestoral Beings.

For Indigenous people who have an interest in land and who are likely to engage with resource development, two fundamental contradictions arise. First, many Indigenous groups have had to fight hard to reclaim their land under Australian law. In general, this has placed a legal burden on Indigenous claimants to demonstrate spiritual attachment or right to forage over the land claimed (under the ALRA) or continuity of customary practice and connection with claimed native title land. Such continuity of customary practice is clearly incompatible with mining because mining disrupts connection and is incompatible with the maintenance of custom. Second, is the issue of sustainable livelihoods. It is broadly unclear how Indigenous people are to retain sustainable livelihood futures after mine closure. Non-Indigenous (or non-local Indigenous) mine workers might simply migrate to the next mine, but for Indigenous people living on their ancestral lands this might not be an option. Paradoxically, the greater Indigenous integration into the mine economy the greater is the likelihood of localised livelihood vulnerability after closure; and the greater the vulnerability to loss of cultural connection to the land that is predicated on living on country. As Bridge (2004) illustrates graphically with reference to mineral-to-waste ratios, large scale mineral producers often operate at environmental regulatory limits. Pollution, even as weak chemically-active waste streams, can threaten livelihoods that are dependent on wildlife harvesting.

Development outcomes

Under the enormous contemporary pressures exerted by the state to homogenise and mainstream approaches to Indigenous development, any divisions within the Indigenous domain are exploited often to the detriment of Indigenous bargaining power. The outcomes from interactions between miners and Indigenous people mediated by the state can be understood from two perspectives: mainstream criteria and local perceptions.

The development outcomes from engagement between miners and Indigenous peoples have been highly variable and dependent on many factors including regional histories of colonisation, the nature of mines, the value of negotiated benefits packages, and the forms of Indigenous engagement with the mine economy. As suggested above, notions of development are culturally constructed and never easy to objectively measure. In the general absence of comprehensive frameworks to independently monitor development outcomes, any study is generally limited to either using official statistics that have not been purpose designed to socioeconomic impact assessment or to undertaking case studies (see Taylor, Chapter 3). A problem with the former approach is that it requires implicit acceptance of the dominant modernisation framework and leaves no room for local perspectives; data are generated from standard census instruments. A problem with the latter is that it can be difficult to make comparisons and to move from the particular to the general. Both approaches are briefly examined here.

Formal outcomes and social indicators

The formal approach in this research project used 2001 Census data for cross-sectional comparative purposes. These data were made available to the project in 2006 at an unusual level of geographic disaggregation. The analysis examined health, education, income, employment and housing social indicators as collected in the census. Eight remote regions with major mines were statistically clustered together.[16] The analysis undertaken in Table 2.1 compares the socioeconomic status of Indigenous people in this mining aggregation with a number of other aggregations: people in adjacent regions, people in remote areas with major tourism destinations, control areas in remote locations where commercial opportunity is largely absent, and then all Indigenous and non-Indigenous Australians. This suggests that there are positive statistical outcomes for Indigenous people from mining activity. In Table 2.2 this socioeconomic information is provided at the level of individual regions within major mine sites. These are the data that were aggregated in Table 2.1. A major qualification for this analysis is that it does not take into account the possibility that in some cases other significant opportunities besides mining might be available in the eight selected regions.

There are indications from Table 2.1 that mining does make a difference in outcomes, according to the statistical social indicators available, although clearly the economic status of Indigenous people in mining areas does not approach that of non-Indigenous Australians. Furthermore, at the individual mine site level there are clear indications that employed Indigenous people do not do as well as non-Indigenous mine workers (ABS 2004). Information in Table 2.2 suggests that there is enormous variability between mining regions.

The social indicators in these tables raise two important issues. First, they demonstrate how badly off Indigenous people are compared to other Australians according to mainstream criteria. This, as we shall see below, raises concerns about whether mining strategies can 'close the gaps' in socioeconomic outcomes. Second, the tables demonstrate how difficult it is to generate appropriate census geographies that can actually measure the impact of major resource developments on host or adjacent communities using official data (see Taylor, Chapter 3). In reality, the impacts on supposed beneficiaries can only be assessed at the local level on a case by case basis. But even at this level, identifying who has benefited can be extremely problematic.

---

[16]   The eight regions are: Gove/Groote Eylandt/Jabiru/East Kimberley/West Cape York/ Borroloola/Gulf/Pilbara respectively operated by Alcoa/BHP Billiton (GEMCO)/Rio Tinto (ERA)/Rio Tinto (Argyle Diamond Mine)/Rio Tinto (Comalco)/Xstrata/Zinifex/Rio Tinto (Pilbara Iron), BHP Billiton, Woodside.

Table 2.1 Social indicator outcomes at mine site and some other jurisdictions, 2001 Census

Demography/health status[a]	Aged under 55	Aged 55 years +	Aged 55 years + (%)
Mining areas	16 306	1 255	7.2
Control areas: Tourism	3 522	287	7.5
Control areas: Other	2 464	183	6.9
Australia: Indigenous	382 420	27 583	6.7
Australia: Non-Indigenous	13 717 012	3 874 477	22.0
Education status	**Not completed Year 10**	**Completed Year 10 or higher**	**Completed Year 10 or higher (%)**
Mining areas	4 694	5 037	51.8
Control areas: Tourism	1 313	759	36.6
Control areas: Other	2 464	183	6.9
Australia: Indigenous	83 616	131 933	61.2
Australia: Non-Indigenous	2 526 996	10 353 748	80.4
Income status			**Median income**
Mining areas			$120-$199
Control areas: Tourism			$120-$199
Control areas: Other			$120-$199
Australia: Indigenous			$200-$299
Australia: Non-Indigenous			$300-$399
Employment rate (aged 15 +)	**Not employed**	**Employed**	**Employment to population ratio**
Mining areas	5 576	4 953	47.0
Control Areas: Tourism	1 510	783	34.2
Control areas: Other	1 162	521	31.0
Australia: Indigenous	140 466	100 393	41.7
Australia: Non-Indigenous	5 689 004	8 144 486	58.9
Home ownership rate	**Not purchasing or does not own home**	**Purchasing or owns home**	**Purchasing or owns home (%)**
Mining areas	2965	461	13.5
Control areas: Tourism	660	82	11.1
Control areas: Other	284	3	1.1
Australia: Indigenous	74 047	37 131	33.4
Australia: Non-Indigenous	1 703 078	4 531 597	72.7

a. Proportion of population aged 55+ years used as a proxy for life expectancy; for further discussion see Altman, Biddle and Hunter 2008.
Source: ABS 2001 Census of Population and Housing.

Table 2.2 Social indicator outcomes in eight mining regions, 2001 Census

Demogrpahy/health status[a]	Aged under 55	Aged 55 years +	Aged 55 years + (%)
Gove/Nhulunbuy	1 207	72	5.6
Groote Eylandt	1 421	85	5.6
Jabiru	467	26	5.3
East Kimberley	487	47	8.8
West Cape York	1 460	136	8.5
Borroloola	1 006	80	7.4
Gulf	5 494	429	7.2
Pilbara	4 466	360	7.5
Education status	**Not completed year 10**	**Completed year 10 or higher**	**Completed year 10 or higher (%)**
Gove/Nhulunbuy	502	244	32.7
Groote Eylandt	666	234	26.1
Jabiru	137	149	52.1
East Kimberley	146	161	52.4
West Cape York	557	427	43.4
Borroloola	403	182	31.1
Gulf	1 193	1 928	61.8
Pilbara	989	1 664	62.7
Income status			**Median income**
Gove/Nhulunbuy			$120–$199
Groote Eylandt			$120–$199
Jabiru			$120–$199
East Kimberley			$200–$399
West Cape York			$120–$199
Borroloola			$120–$199
Gulf			$200–$399
Pilbara			$200–$399
Employment rate (aged 15 +)	**Not employed**	**Employed**	**Employment to population ratio**
Gove/Nhulunbuy	384	410	51.6
Groote Eylandt	628	240	27.7
Jabiru	121	180	59.8
East Kimberley	130	199	60.5
West Cape York	437	595	57.7
Borroloola	251	414	62.3
Gulf	1 840	1 576	46.1
Pilbara	1 698	1 256	42.5
Home ownership rate	**Not purchasing or do not own home**	**Purchasing or own home**	**Purchasing or own home (%)**
Gove/Nhulunbuy	167	3	1.8
Groote Eylandt	227	3	1.3
Jabiru	101	19	15.8
East Kimberley	68	7	9.3
West Cape York	262	0	0.0
Borroloola	167	5	2.9
Gulf	959	250	20.7
Pilbara	963	171	15.1

a. Proportion of population aged 55+ years used as a proxy for life expectancy; for further discussion see Altman, Biddle and Hunter 2008. Source: ABS 2001 Census of Population and Housing.

Local views about outcomes

An analysis of local situations is clearly enormously complicated and an attempt is made here to briefly canvass issues that arise in the three cases, the Ranger Uranium Mine (Northern Territory), the Yandicoogina Iron Ore Mine (Western Australia) and the Century Zinc and Lead Mine (Queensland) (see Fig. 2.1) that are given more coverage in later chapters. In each case major benefit sharing agreements have been completed: in the case of Ranger signed in 1978 under the ALRA; in the case of Yandi as an Indigenous Land Use Agreement completed outside, but informed by, the native title framework in 1997; and in the case of Century under the NTA right to negotiate procedures. These are three landmark and major agreements in the Australian context and a brief examination of each will show that regional studies suggest that outcomes might be far less beneficial than official statistics suggest. By looking at particular agreements it is possible to also dig a little deeper to look at environmental, social, cultural and other impacts that go beyond the criteria dictated by a limited range of mainstream social indicators.

Fig. 2.1 The location of the three case studies

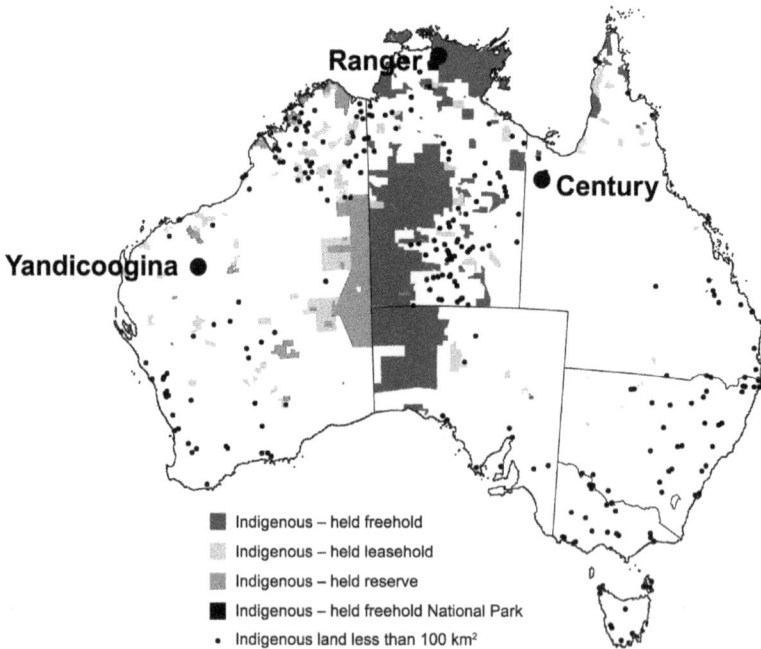

Case 1: The Ranger Uranium Mine agreement

The Ranger Uranium Agreement was the first major post-ALRA mining agreement completed. It has many unusual features: it was a prior interest mining prospect, so the right of veto could not be exercised; the Australian Government was a major stakeholder in the uranium project so was in a conflict of interest as a party both to the mine and the agreement with traditional owners; and the agreement was signed as part of a regional deal that included the creation of Stage 1 of Kakadu National Park and a settled land claim (Fox, Kelleher and Kerr 1977). The Ranger Agreement covered a range of issues including benefit flows, employment at the mine, and environmental protections. In its early days, it was judged to be a very positive agreement. The Gagudju Association, a specially incorporated regional organisation, received 'areas affected' monies. These monies were utilised between 1981 and 1996 to provide services to about 300 members; to make investments, especially in tourism infrastructure and enterprises in Kakadu National Park; and to make limited cash payments to its members. Since 1978, royalty equivalents of over $A200 million have been paid in relation to this mine, with over $A60 million flowing back to the region under the ALRA formula.

The very positive story of the Gagudju Association soured somewhat from the mid-1990s after the traditional owners of the mine site, the Mirarr Gundjeihmi, sought the re-channelling of areas affected monies to a far smaller group of 26 adult traditional owners. The cause of this change was evident in a politically complex dispute that had its genesis in two key issues (Altman 1997). First, the Mirarr were concerned that mining had had a negative impact on local Aboriginal people. Second, and related to this, they opposed the mining of another major uranium prospect, Jabiluka, that has lain dormant since 1983 owing to the Commonwealth Labor Government's 1983–96 freeze on new uranium mines. A subsequent comprehensive social impact study, the Kakadu Region Social Impact Study (1997b, 1997c) undertaken by two groups[17] broadly supported traditional owner concerns. It was noted, for example, that mining payments had generated regional conflict and were largely offset by reduced state expenditure in the provision of normal citizenship services on a needs basis. A study using available social indicators demonstrated that Aboriginal people in the mine catchments were no better off than people in adjoining regions (Taylor 1999).

A recent report, 'Aborigines and Uranium: Monitoring and Health Hazards' (Tatz et al. 2006), documents high cancer rates in the region, which has generated vigorous debate about whether these are linked to uranium or not. Since 1981 there have been more than 120 spillages and leaks of contaminated water at the mine; a major spill in 2004 resulted in temporary mine closure and subsequent successful prosecution by the Northern Territory Government in 2005 for breach of environmental regulations. The predictions of the Fox Inquiry in 1977 that mining in the region could have negative social impacts appear to have been

---

[17] The Kakadu Region Social Impact Study was undertaken by a mainly local Aboriginal Project Committee (1997b) and a mainly external Study Advisory Committee (1997c). I was the independent expert to the latter.

prescient. There is little evidence of sustainable economic benefits from the mine whose life has now been extended to beyond 2020 owing to an increase in the price of uranium oxide in recent years. Local Aboriginal people are reluctant to work at the mine; those who are 'job ready' prefer instead to take up jobs in the national park or in tourism, although the unemployment rate for locals remains high.

Case 2: The Yandicoogina Land Use Agreement (YLUA)

The YLUA was the first major post-NTA agreement, but it was completed as a negotiated agreement outside the NTA framework. The signatories were Hamersley Iron (now Pilbara Iron, a wholly owned subsidiary of Rio Tinto) and the Gumala Aboriginal Corporation representing 430 Nyiyapareli, Banyjima and Yinhawangka people. The agreement covers an area of approximately 26 000 square kilometres. Initially the agreement was to provide $A60 million to the Gumala Aboriginal Corporation over 20 years, but a scaling up of production at Pilbara Iron's Yandi mine and a confidential payments schedule might see payments ramped up, possibly over a shorter period.[18]

The YLUA was completed in the Pilbara, a region where Aboriginal people had historically been actively involved in mining on a small 'cottage' scale until driven off by large scale development and state policies in the 1960s (Holcombe 2006). As noted earlier, Western Australia is the only jurisdiction in Australia that has no land rights law and the State government has always taken a very pro-mining position, often in the face of significant Aboriginal opposition. The YLUA was seen positively a decade ago as possibly being the harbinger of a new era in relations between mining companies and Indigenous people: its focus is on the provision of jobs, training, and business enterprise opportunities for local people. A major element in the agreement is site clearances which are required under the Western Australia Aboriginal Heritage Act for broad acre open cut iron ore extraction; this weak statute requires that sites of cultural significance be recorded and that any movable cultural objects be relocated. Rather than contracting heritage clearance work to an independent third party, Pilbara Iron manages this process in collaboration with Aboriginal consultants in a manner that could be perceived as problematic (Holcombe 2006). This situation has evolved for historical reasons when Indigenous representative organisations were not operating effectively in the region.

A recent comprehensive study by Taylor and Scambary (2005) analysed regional statistical social indicators, while also seeking detailed Indigenous views about development impacts. On the statistical side it was shown that employment status for Indigenous people had barely changed for 30 years despite

---

[18] It is noteworthy that the extensive Yandi deposit is mined by both Rio Tinto-owned Pilbara Iron and BHP Billiton who have a separate agreement with a similar group of native title interests.

multinational corporations like Rio Tinto and BHP Billiton setting ambitious employment targets for Indigenous employment. The authors found that, despite strong demand for Indigenous labour, there were significant problems on the supply side owing to poor education, health, substance abuse and high interaction with the criminal justice system. Much of this was linked to state neglect over the past 40 years. From the Indigenous perspective, there was disappointment that the YLUA appeared focused primarily on mainstream outcomes associated with engagement with the Yandi mine, rather than in meeting the diverse aspirations of local people that included living at small outstation communities and engaging in non-mainstream economic activity. In particular, there was concern that the YLUA tied up compensatory payments in trusts that could not be accessed by supposed beneficiaries. Scambary (2007) summarises this in a widely articulated local sentiment: 'We've got the richest trusts but the poorest people'.

Case 3: The Century Mine agreement

This agreement is generally referred to as the Gulf Communities Agreement (GCA). The GCA was signed in 1997 between Century Zinc Limited (CZL) (then a subsidiary of Rio Tinto, now owned by Zinifex), the Queensland Government and the Waanyi, Mingginda, Gkuthaarn and Kukadj people after prolonged and at times bitter political dispute (Martin 1998a). The agreement was signed under the NTA framework before the diluting 1998 amendments and covers the Century zinc/lead mine, a 350 kilometre slurry pipeline, and a port facility.

The GCA provides $A60 million over 20 years to signatory groups, but unlike the YLUA also includes significant State government commitments of an anticipated $A30 million for the provision of education and training, infrastructure and the conduct of a major regional social impact assessment. It is noteworthy that the leverage provided by the NTA right to negotiate process under s.29 saw an initial CZL offer of $70 000 in cash increased to an eventual $A60 million agreement package (Trebeck 2007a). The GCA is a little unclear about the intended beneficiaries of the agreement: reference is made to the provision of employment opportunities at the mine to an estimated 6,000 Aboriginal people in the southern Gulf of Carpentaria while cash benefits are limited to an estimated 900 members of the four signatory language groups.

The Century Mine has been one of Australia's most successful in employing Indigenous labour, with over 100 people or consistently about 20 per cent of the mine site labour force being Indigenous (Barker and Brereton 2004, 2005). This is an undoubted success of the GCA. However, the GCA probably over-ambitiously seeks to remove a large and diverse Indigenous population in the region from welfare dependency and to promote economic self sufficiency—admirable goals that one mine agreement cannot deliver. A five-year review of the GCA in 2002 by a number of parties to the GCA recognised this and highlighted problems in a number of areas, including establishment of viable organisations to channel compensation payments to beneficiaries.[19] Subsequent local dissatisfaction with the review process and with the company's approach to community relations saw a 'civil disobedience' sit-in at the mine that threatened its operations: as Trebeck (2007a, this volume Chapter 6) notes, the sit-in demonstrated that local communities had the potential to impede operations. Subsequently, a separate review of the GCA by the regional representative body articulated community concerns especially in relation to meeting obligations in the GCA to fund outstation and other development (Carpentaria Land Council Aboriginal Corporation 2004). These criticisms were directed as much at the Queensland State as the mining company, but the political lever was far greater with the latter. As in the other two cases, there are tensions following the GCA between Indigenous beneficiaries groups about distribution of benefit payments and a lack of clarity about intended agreement beneficiaries especially with respect to distribution of employment and training opportunities (Scambary 2007).

The three cases briefly canvassed here highlight six commonalities that can be summarised as follows:

In all cases there are expectations mismatches that reflect differential power relations and an inability of agreements to recognise regional Indigenous diversity. Clearly some Indigenous people see mines as providing opportunity for formal employment while others see agreements as a means to pursue life projects (Blaser 2004; Peterson 2005; Trigger 2005). Some Indigenous people interviewed by Taylor and Scambary (2005) articulated a clear aspiration to use agreement benefits to live on their land and engage in the customary sector rather than in mine site employment. From the Indigenous perspective, there was little indication that mining agreements were investing in sustainable futures.

It is unclear who the intended beneficiaries of agreements are. On one hand, supposed benefits are in the nature of mine site employment that also benefits the state and the mining industry. It is very clear in all situations that intended Indigenous beneficiaries have limited autonomous control of how benefits are spent and there is limited capacity for adaptive management of agreements (Ballard and Banks 2003).

Demarcating government from company responsibilities is extremely problematic. Historical underinvestment in social and physical infrastructure by the state means that many Indigenous people do not have the capabilities to work at mines, even if they wished to. There is a tension between mining companies and the state about who should deliver basic services and infrastructure with some evidence of cost shifting from the state to mining companies (Altman 1983;

---

[19] It should be noted that the author was engaged as an adviser to this review.

O'Faircheallaigh 2004a). There is recent evidence that Indigenous peoples have articulated a desire to sign mining agreements with companies to gain access to essential services that should be provided by the state.[20]

Friction within the Indigenous population of mine hinterlands is common, with major tensions arising from lack of clarity about intended beneficiaries and consequent political conflict between traditional owners (who may or may not live near the mine) and other Indigenous people, some of whom have been long-term regional residents (so called historical people) and some who may have migrated specifically for mine site employment. Mine site statistics show great variability in local and non-local Indigenous employment (Tiplady and Barclay 2007).

The capacity of Indigenous people to benefit from major resource developments, even where they seek engagement, can be extremely limited owing to a legacy from past neglect of poor health and education and high arrest rates that have made them unsuitable for mine employment. A particular problem that Indigenous peoples face is poverty traps (represented by extremely high effective marginal tax rates, sometimes over 100 per cent) that undermine incentives for mine site employment. These can be exacerbated by agreement payments to individuals, although these are rare.

Everywhere there are environmental concerns. Examples from the cases being discussed include the potential to impact on a sacred site or on food-rich wetlands at Jabiluka; the impact of landscape scale strip mining on cultural heritage sites in the Pilbara; or on coastal subsistence fisheries or the location of a cyclone-mooring buoy on a sacred site in the Gulf of Carpentaria. These environmental concerns can provide fruitful bases for Indigenous alliances with environmental non-government organisations as occurred very effectively at Jabiluka (Trebeck 2007a, this volume Chapter 6; Triggs 2002).

Contestation over development

In historical terms the past decade or so is a period when the state has clearly moved to dilute Indigenous rights, reflecting a particular political climate and a national focus on economic imperatives, as articulated by the elected government of the time. A political economy analysis might suggest that the state and capital have formed an enduring alliance that marginalises Indigenous people. Yet even such an analysis raises questions about enlightened self interest (Ferguson 1994) that would suggest that the state's economic growth project might be best served by addressing Indigenous poverty and marginality in mine hinterlands specifically and Australian society generally.

---

[20] R. Taylor, 'Aboriginals say uranium mines answer to poverty', *The West Australian*, 29 June 2006, p. 6.

Bridge (2004: 205) suggests that four distinctive approaches (technology and management centered accounts, public policy studies, structural political economy and cultural studies) can be used to address this question. Similarly a large number of analytical approaches could be used to understand the contestation over development evident in remote Australia between mining companies, the state and Indigenous peoples. I focus here on two broad approaches, political economy and cultural analysis, to highlight power differentials that are seeing the state undertaking a strategy to depoliticise Indigenous institutions and to incorporate Indigenous people in a monolithic project of industrialisation; simultaneously, and against the odds, Indigenous people are mobilising to highlight cultural difference as a means to articulate their diverse and different notions of development in relation to their lands.

The global discourse of development today is dominated by the perspectives of affluent states and multinational corporations. Despite growing global concerns about climate change, resource depletion and limits to growth, economic liberalism remains in the ascendancy (Harvey 2007). And despite occasional nods to multi-cultural citizenship (Kymlicka 1995), the power of states to define development trajectories for its citizens is growing. This is despite alternate views that, with globalisation, the nation state will lose influence to multinational capital (Blaser 2004; Howitt, Connell and Hirsch 1996).

In the Australian context, economic liberalism and its associated discourse of development is in the ascendancy, recognition of special Indigenous rights is currently at a low point (despite the new Rudd Government supporting the United Nations Declaration on the Rights of Indigenous Peoples), and the state's power to define a development trajectory for the Indigenous minority is hardly challenged in popular or policy discourses or by high profile Indigenous leaders some of whom actively advocate for this project of improvement. The state appears all powerful, immune from serious international scrutiny of its domestic policies—perhaps because the global community itself is too eager for access to Australian minerals?

State power, with a compliant media, is increasingly exercised to define development in terms that reflect dominant group values. Somewhat paternalistically, after abolishing the Aboriginal and Torres Strait Islander Commission (the national Indigenous representative organization) in 2004, the state is limiting its notion of Indigenous development to mainstream aspirations: employment success, high monetary incomes, individual home ownership, entrepreneurship and material accumulation. This somewhat hegemonic and monolithic take on development is problematic for many Indigenous groups and especially those living on the Indigenous estate in remote regions. It is based on universalism, a focus on the individual, a growing intolerance of cultural difference, and a limited view of development that is committed to market-based solutions to deeply entrenched Indigenous marginalisation.

There are two broad readings of this approach each with its own inherent contradictions and inconsistencies. The first is that the political and bureaucratic elites genuinely believe that mainstream development at mine sites can deliver regional socioeconomic statistical equality. Such a policy aspiration makes some sense if 'seeing like a state' (Scott 1998) for it could reduce the high direct and indirect costs to the state of Indigenous dependency, generate labour in situations of shortage, and generate additional national wealth from taxation of Indigenous workers.

The second equally plausible reading is that the extent of Indigenous land ownership that has incrementally grown over the past three decades as a result of new laws based on social justice principles and new legal interpretations, is just too great. It is certainly the case that land rights and native title laws have seen the Indigenous estate grow to over 20 per cent of the Australian continent. While the state might not seek to openly dispossess Indigenous people of their new land holdings, it might seek to facilitate access to such lands for mineral exploration and extraction by weakening resource rights, weakening associated Indigenous negotiating power, and weakening the Indigenous institutions that can effectively negotiate on behalf of their constituents.

The 'mainstreaming for engagement' versus the 'mainstreaming for exploitation' readings have logical contradictions. Attempts at Closing the Gap(s) would benefit from a strengthening not weakening of Indigenous property rights, while exploitation of the Indigenous estate is hardly likely to close the gaps unless Indigenous people embrace mine employment. Similarly, it is the extent of Indigenous dependence on the state that leaves them so vulnerable to state interventions—'welfare poison' can disempower (Pearson 2000a). Hence while the state's particular form of development demands enhanced Indigenous engagement with mining, the social investments to facilitate such participation are inadequate and the property rights frameworks that might enhance Indigenous levers to negotiate for such investments are weak.

Mining companies are keen to deal directly with traditional owners to gain social licence to operate on the Indigenous estate. Owing to security concerns, the massive 'greenfields' Indigenous estate within the strong Australian state is highly desirable to multinational corporations. Direct relations between miners and Indigenous peoples are clearly on an upward trajectory, as evidenced by the hundreds of agreements referred to by the MCA (2006). Two recent studies (Langton et al. 2004, 2006) clearly show that agreement making is on the rise.

As material on the MCA website shows, its members (who include some of the world's most powerful multinationals) have embraced corporate social responsibility as a new approach in relations with Indigenous communities.

However, as Trebeck (2005, 2007a, this volume Chapter 6) notes in her study of the Yandi, Ranger and Century mines, while head office might hold lofty responsibility ideals and sponsor affirmative programs, these ideals are not necessarily penetrating to mine site business units. Not only do these units have greater capacity for technical excellence than social policy, but the financial bottom line looms large as a priority for mine managers. There are clear tensions between social, economic and environmental elements of the triple bottom line.

At times these tensions erupt into civil disturbances at or near mine sites and individual mining companies have not been immune from such Indigenous activism at Marandoo (in the Pilbara), at the Jabiluka prospect, and at the Century mine. While sovereign risk in Australia is low, from the Indigenous perspective such activism provides a means to demonstrate local dissatisfaction with mining activity and to leverage and then re-negotiate for post-agreement mine site changes. At Jabiluka, Indigenous traditional owners scored a rare victory when, via environmental non-government organisation alliances, shareholder activism and the prolonged and zealous campaigning of the Gundjeihmi Aboriginal Corporation (that included visits to the United Nations Educational, Scientific and Cultural Organisation (UNESCO) in Paris resulting in the UNESCO Kakadu Mission in 1998[21] and to the Rio Tinto Annual General Meeting in London), the company in 2000 agreed not to mine Jabiluka without traditional owner consent. This outcome is iconic because it represents the only occasion in Australia where traditional owners have seen the rescinding of an existing agreement (signed under duress between the Northern Land Council, Pancontinental and the Australian Government in 1982).[22] Very importantly, this decision by Rio Tinto has been taken despite state pressures to proceed with mining.

Just as individual mining companies have operated more responsively to traditional owner views than the state, so too the MCA has repositioned itself. In the past, the Council used to campaign for amendments to the law to dilute Indigenous political and economic leverage. More recently, as evidenced by its recent submission to the Commonwealth government (as well as to Senate Inquiries into proposed amendments to land rights law in 2006 and to the NTA in 2007), it has begun to campaign for enhanced government resourcing of remote Indigenous communities and for support for Indigenous representative organisations that have a degree of independence and capacity. This is a surprising development that says as much about the disempowering

---

[21] I was an Australian government appointed member of the Mission.

[22] ERA could have mined Jabiluka but would have needed to process the uranium ore on site, something that was not financially viable in the late 1990s. Consequently, ERA needed to gain traditional owner consent to transport ore to the Ranger milling facility some 25 kilometres away, permission that was not forthcoming from the Mirarr Gundjeihmi owners of the transport corridor. It is paradoxical perhaps that the anti-Jabiluka campaign was at least part-sponsored by an organisation set up with Ranger Uranium Agreement payments. However, arguably, institutional capacity must be resourced from somewhere to provide independent voice.

mainstreaming measures of the Australian Government as it does about the new ways of doing mining business in Australia.

Indigenous views on development can be very different from mainstream notions. As already noted this is especially the case in remote regions where the struggle to win back land has often required provision of proof in the courts of land connection and extant customary practice. Having won back legal title to land, many groups are frustrated that their traditional lands are nevertheless available for exploration and mining in all jurisdictions, except where there are free prior informed consent provisions. Owing to their minority status and limited political power, Indigenous views about development form a subordinate discourse that has great difficulty being heard.

Peterson (2005) uses recent writings from North America (Blaser 2004) to distinguish Indigenous 'life projects' from state 'development projects'. Such life projects are structured by colonial history and ongoing relations of high dependence with the state. They are also structured by ongoing engagement in the customary (non-market) land-based sector of the economy. Trigger (2005) assesses mining projects in remote Australia as sites for contestation between Indigenous and mainstream views about economic and cultural futures. These contestations can be stylised as a clash between market-based and kin-based economies (Austin-Broos and MacDonald 2005) and much more. As a general rule, it is clear that Indigenous land owners look to maintain the environmental integrity of their land, whilst miners look to exploit the land's non-renewable resources; Indigenous people see the land and the landscape as a cultural asset, not just a commercial asset. An important strength of Trigger's analysis is his engagement with mining as an intercultural process: his research reports that some Indigenous people believe that they can commercially engage with mining, while maintaining their identity and distinct cultural practices. To paraphrase Stanner's poignant words, there are some Indigenous people who believe that the market and the Dreaming are compatible, others who do not.[23]

The problem that Indigenous people face is that in the face of a power narrative of Indigenous policy failure, there is a growing national intolerance of Indigenous diversity and cultural differences. This is very clearly encapsulated in the couching of the dominant development discourse in terms of practical reconciliation or statistical equality between Indigenous and other Australians. Such policy focus on Closing the Gap has been rejected by Indigenous people elsewhere internationally because it pathologises Indigenous disadvantage by defining it in relational terms to mainstream standards that are constructed according to distinct non-Indigenous cultural values (Durie 2005; Smith 1999; Storey 2003). In this way very different Indigenous views of development are

---

[23] Stanner's exact words were 'Ours is a market-civilisation, theirs not. Indeed there is a sense in which The Dreaming and The Market are mutually exclusive' (1979: 58).

marginalised and development debates are structured by the dominant western paradigm.

The capacity of Indigenous peoples to resist state-sanctioned mining or to ensure equitable benefit sharing is highly dependent on enhancing the capacity of regionally-based Indigenous organisations to protect Indigenous legal rights. The current approach using existing organisations and negotiation levers is resulting in Indigenous people becoming increasingly vulnerable to unequal agreement making (see Corbett and O'Faircheallaigh 2006) and increasingly being unable to oppose development pressures.[24] This is a growing problem because in recent years the state has become increasingly intolerant of dissenting views, be they in the academy, the public service, or in advocacy support (Hamilton and Maddison 2007). In some cases, as documented by Trebeck (Chapter 6), local Indigenous agency and activism has clearly demonstrated the vulnerability of mining companies to local hostility. This might explain why companies are more responsive than the state to Indigenous perspectives and why the industry is concerned about inequities in bargaining power. An adversarial approach, however, may not be sustainable as a source of bargaining strength.

Reconciling different views of development

This chapter has focused on three very broad categories—Indigenous peoples, states and mining companies—and examined their interactions in development processes. As with all such articulations there are clearly category overlaps and cleavages: Indigenous people are citizens of Australia and companies operate within the nation's borders and in accord with Australian laws. In terms of motivating simplifications, it could be argued that the state seeks authority and compliance with its dominant notion of development, Indigenous peoples seek autonomy and the right of self governance, and mining companies seek licence to operate and secure access to resources.

In terms of dominant ideology, it is increasingly the case that the states subscribe to economic liberalism, while Indigenous peoples are seeking recognition of their right to be different and to be heard—as outlined in post-development theory that problematises western notions of development (Storey 2003: 34–37). In Australia, this tension is evident in debates about practical versus symbolic reconciliation, economic equality versus cultural plurality, and market-based versus kin-based economic systems. Is some commensurability between such binaries a possibility, or are what Mander and Tauli-Corpus (2006) term 'paradigm wars' inevitable?

---

[24] Corbett and O'Faircheallaigh (2006) argue that even when using the available and supposedly impartial National Native Title Tribunal institutions for arbitration, native title parties are disadvantaged and 'grantees' (mining companies) advantaged.

An alternative model that is applicable to Indigenous people in remote Australia is the hybrid economy framework (Altman 2005a). This model is based on a critique of orthodox development approaches that privilege the market and the state and ignore the customary or non-market sector of local economies. The model has some commonalities with both the livelihoods (de Haan and Zoomers 2005) and community economy approaches (Gibson-Graham 2005).

The hybrid economy model is depicted conceptually and diagrammatically in Fig. 2.2. To simplify considerably, it is made up of three sectors (represented by the circles marked 1, 2 and 3). A crucial feature of the model is the articulations (or inter-linkages) between these sectors (depicted by the segments 4, 5, 6 and 7). An important feature of the model is that the relative scale of the three sectors and four points of articulation vary from one context to another. In remote Australia, many Indigenous people regularly move between these seven segments with the mobility evident in pre-colonial times in the food quest now evident in livelihood adaptations. For example, an individual might participate in wildlife harvesting for domestic use, the production of an artefact for sale, employment at a mine site or in the public sector or be in receipt of income support from the state.

Fig. 2.2 The hybrid economy framework

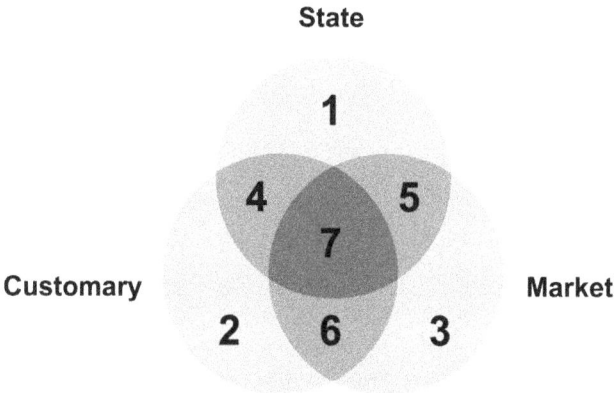

In such circumstances people are not solely reliant on welfare, on the non-market sector, or on income from market engagement. In a sense, part of the emerging post-colonial adaptation observed is a risk-minimisation strategy, whereby a diversity of sources of livelihood results in engagement in all sectors of the local economy. What differentiates the Indigenous Australian situation from many other Third World situations is the central role of the state providing citizen entitlements to various degrees. This support occurs directly, for example, in the provision of income support and indirectly, for example, through the provision of state patronage of community enterprise.

While the notion of the hybrid economy in Australia grew from case study work in Arnhem Land among an encapsulated minority in a post-colonial state who have shaped their local economy into a very distinct form, there are two broad reasons to believe that this concept has wider Australian applicability.

First, a major survey undertaken by the Australian Bureau of Statistics, the National Aboriginal and Torres Strait Islander Survey 2002 indicated a very high level of Indigenous participation in non-market activity in hunting and fishing and in paid cultural activities like production of art for sale (Altman, Buchanan and Biddle 2006). While the 2002 Survey had major shortcomings in fully capturing the significance of all elements of the hybrid economy, it certainly demonstrated that the land had important productive value for Indigenous people which is often overlooked. It also reinforced the view that the nature of the economic problem in remote Indigenous Australia is misunderstood as the level of engagement with the customary sector is overlooked.

Second, in Australia as elsewhere, climate change and associated national concerns about water quantity and quality and potential loss of biodiversity are all high priorities. Recent research shows that the Indigenous estate includes some of the most biodiverse lands in Australia (Altman, Buchanan and Larsen 2007). Official natural resource atlas maps produced by Land and Water Australia and the Department of Environment and Heritage indicate that many of the most intact and nationally-important wetlands, riparian zones, forests and rivers and waterways are located on the Indigenous estate. Mapping also shows that these lands are at risk of species contraction and face major threats from feral animals, exotic weeds, changed fire regimes, pollution and overgrazing. Potentially there is a crucial role for Indigenous people in environmental management of the Indigenous estate they own. This is an area where Indigenous ecological knowledge and Western science can be linked and where Indigenous people are actively seeking enhanced engagement. While much is already undertaken, Indigenous people are poorly remunerated for the provision of a range of environmental services. There are significant opportunities to enhance such Indigenous engagement as an element of the hybrid economy that could be supported by the state or by mining companies in environmentally managing their properties.

The hybrid economy model does not seek to ignore the contestation within the Indigenous domain about economic futures. There are influential Indigenous leaders (as noted by Trigger 2005) who advocate a fuller Indigenous embrace of the market as a means to create wealth and long-term prosperity and address social problems. In the hybrid economy model, market engagement is not precluded. As noted earlier, some people near mine sites seek employment with various motivations including as a means to accumulate financial resources to get back onto their country (Scambary 2007, this volume Chapter 9). This highlights the plurality of development approaches sought by Indigenous people. The question remains, can the state and mining companies accommodate and foster such plurality while also facilitating productive engagement with mine economies?

It is here that contradictions emerge, because even within the state there is debate—evident, for example, in the very different approaches of industry and environmental agencies. Dominant political and bureaucratic views support expansion of mining onto the Indigenous estate, while less powerful environmental agencies are seeking to include more and more of the Indigenous estate into the conservation estate using the Indigenous Protected Areas program. On the other hand, mining companies appear reluctant to directly support employment in natural resource management, choosing instead, perhaps from self-interest, to focus all effort on training and employment on mine sites even if take up is poor. There may be more productive ways to engage with Aboriginal people in mine hinterlands especially given the extent of occupational migration between all sectors of the 'hybrid' economy.

Conclusion

This chapter highlights contestation over economic development, with the state and mining companies looking to exploit the mineral wealth of the Indigenous estate, and Indigenous Australians often seeking to challenge this powerful alliance with different and diverse notions of development. It has been argued that the state, having made laws in the recent past to return considerable tracts of land to Indigenous Australians, is now looking at this estate as 'greenfields' for mining development and as the means to deliver mainstream development to Indigenous communities. The state has constructed a new discourse of policy failure and a need to break business-as-usual approaches. Arguably, the state is also seeking to encourage commercial penetration of the Indigenous estate by weakening already weak Indigenous property rights and institutions in a manner that is at odds with emerging international conventions on the recognition of Indigenous rights such as numerous articles with the United Nations Declaration on the Rights of Indigenous Peoples.

The analysis here highlights two fundamentally different views about the appropriate development pathways on the Indigenous estate. The perspective of dominant national interest and global geopolitics seeks to explore and exploit the mineral wealth of the Indigenous estate irrespective of the wishes of the land owners. An alternative view, based on emerging alliances between Indigenous land owners and environmental and developmental non-government organisations, emphasises the biodiversity value of the Indigenous estate as an alternative form of development to mineral extraction.

At present, the former perspective is in the ascendancy, dominated by the views and political and economic power of the state and mining companies. At the start of the twenty-first century there has been an acceleration of a new economic order predicated on world trade and energy-intensive industrialisation that is right now being challenged by a global slowdown. As a commodity-export dependent economy, Australia has been at the vanguard of the neoliberal order that has been so dominant in recent years. At the same time, there are concerns about climate change that the Australian Government, in recent years, has been at the vanguard of ignoring, at least until a change of national government in November 2007. At such a time it is extremely difficult for any alternative development perspective, based on proven links to land and continuity of custom, to gain political traction. This is especially the case because the Australian state is in the process of depoliticising Indigenous institutions, and mainstream political channels reflect the views of the majority only.

The means that the powerful use to support arguments in favour of modernisation is to suggest that Indigenous people living in remote Australia face a stark choice between tradition and modernity: the former is associated with material poverty, the latter with affluence. This is a false dichotomy because Indigenous people in remote Australia are already participating in a hybrid economy that is thoroughly intercultural and is inclusive of both the customary and the market. Indigenous people undergo rigorous tests to show that they maintain custom and identity in order to regain land ownership; they now face new challenges in retaining that land ownership. There are immense pressures to join the mainstream, but the issue then arises of whether economic integration is possible for Indigenous Australians without losing their links to the land. Conversely, if a choice is made not to join the mainstream, then there is the unenviable prospect of ongoing poverty and marginalisation.

There are many possibilities that are currently not being considered. A new approach is needed that gives institutional recognition to the inherent rights of Indigenous Australians to control the nature of development on their ancestral lands by bestowing them with more potent property rights. Indigenous peoples need to be empowered by the state, as noted by the MCA, so that they can participate in negotiations with mining companies on a more equitable basis. Such empowerment will ensure that Indigenous Australians have the capabilities and capacities to engage with the state and mining companies, and can define their own aspirations. This is a right, as espoused by Sen (1999), to develop capabilities to negotiate the forms that development and participation will take. The Australian states and mining companies need to recognise that the security of the poor and of the prosperous are inter-linked; poverty and instability has the capacity to destabilise states and to enhance risk.

In the political struggle over ideas about development and how Indigenous land should be used, Indigenous Australians need to use available political instruments and alliances to gain a hearing. This may entail appeal to international forums, although as noted the Australian state has historically been relatively impervious to international opinion. There is a glimmer of hope in the shifting national mood and changes in direction of the moment (post the onset of the global financial crisis) that might see a greater acceptance of alternate development models, like the hybrid economy framework, that have the capacity to incorporate Indigenous priorities. The challenge for the Australian state is to recognise the value of cultural diversity and economic hybridity; and to acknowledge that there is nothing inherently valuable about monolithic approaches based on resource extraction. The challenge for mining companies is to recognise that the fostering of sound regional relations with Indigenous people might involve agreement making that supports their aspirations for economic diversity beyond mine site options. Given poor past relations, Indigenous scepticism about rampant development is understandable. Whether the marginal situation of Indigenous peoples in remote Australia can be improved will be highly dependent on acceptance by the state and mining companies that all their interests are interdependent.

3. Data mining: Indigenous peoples, applied demography and the resource extraction industry

John Taylor

In recent decades, applied demography has emerged as a sub-field of demography partly in response to a growing demand from governments and private sector interests to better understand the implications of population trends for public policy and business strategy (Murdock and Ellis 1991; Siegel 2002). While in essence, this involves the practical application of demographic materials and methods (Siegel 2002: 2), the emphasis is on gaining knowledge of the consequences and concomitants of change in the size of populations, their distribution, composition and characteristics, so as to guide decision-making related to planning and the distribution of public or private sector goods and services for current and future use (Murdock and Ellis 1991: 6). This is precisely the empirical input that industry, government and Indigenous stakeholders have begun to identify as contributing to meaningful discussions about options for integrating the activities of mining operations with broader social and economic development goals in mine hinterlands. Accordingly, a push for profiling regional social and economic conditions has emerged from a coalition of these interests (Harvey and Brereton 2005).

Underlying this push, the current political economy of minerals development across remote Australia attempts to ensure that Indigenous peoples and communities increase their capacity to participate in bouyant regional economies that are stimulated by the 'super-cycle' of global demand for mineral products (Harvey 2002a; Hooke 2007). Overall, the aim is to establish initiatives that will secure sustainable economies for mining regions beyond the operational life of mines, making full use of local labour (Harvey 2002a, 2004). From a minerals sector perspective, this reflects the growing influence of corporate social responsibility and a corresponding recognition of necessary foundations for a social licence to operate (Trebeck, Chapter 6). From an Indigenous perspective, it reflects the assertion of a legitimate stake in directing regional development options, not least on Aboriginal lands. And from a government perspective, it reflects a window of opportunity to realise policy goals of Closing the Gap via the activities and economies of scale induced by minerals development.

Measuring impacts

One way in which these varied aspirations are pursued is through negotiated mining and other region-specific agreements. In the early 1970s, there was considerable optimism that mining agreements, many with significant financial benefit packages, would make a difference to the marginal economic situation of Indigenous beneficiaries (Altman 1983). However, research to date indicates that for a complex set of reasons, Indigenous economic status has changed little in subsequent years—dependence on government transfer payments across remote Australia remains high, while the economic profile of Indigenous people residing in the vicinity of major long-life mines is often indistinguishable from that of Indigenous people elsewhere in remote Australia (Taylor 1999, 2003; Taylor and Scambary 2005).

This situation partly reflects the incapacity of Indigenous community organisations and individuals to cope with the impacts of large-scale operations and take advantage from them. Equally, though, such organisations and the people they represent may have ambivalent responses to the potential cultural assimilation implied by their increasing integration into a market economy and its associated monetisation of many aspects of social life. A third key factor has been the attitudes and responses of both mining companies and governments, and their inability to comprehend and respond to the scale of historic Aboriginal disadvantage and strain on the social fabric of societies so radically affected by colonisation.

As well as these issues, it is now also clear that the content of many agreements has been far from optimal and that substantial legal, policy and institutional obstacles remain in the way of favourable negotiated terms related to mining activity on Aboriginal lands (Altman, Chapter 2; O'Faircheallaigh 2006). One issue in particular that has compromised the quality (or effectiveness) of agreements, is the general lack of attention paid to initiatives and activities required to give effect to stated provisions. The result, in many instances, is a failure of implementation for want of adequate resources and structures to ensure that this occurs and is sustained (O'Faircheallaigh 2002a).

Viewed historically, this general failure of agreements to monitor progress and impacts may be seen as a lesson once learnt, then subsequently lost, since the prototype for monitoring emerged as a key recommendation of the Fox Inquiry as far back as 1977. This inquiry called for the establishment of a five-year monitoring study of the social impact of uranium mining on Aborigines that was undertaken by the Australian Institute of Aboriginal Studies. While this study openly eschewed what it referred to as the 'technocratic tradition' and the hard-edged statistical approach of the policy sciences, it nonetheless aimed to generate as much baseline data as possible against which to measure social change. In line with a movement towards what was described as a 'political' or 'community development' model of social impact assessment (Ross 1990: 12), this was to be developed as a computer-based store of relevant information that could answer numerous questions for Aboriginal communities, government agencies, mining companies and researchers, and be functional for decades to come (Australian Institute of Aboriginal Studies 1984).

While the scope of this study was sound enough, it is interesting to note that almost 20 years later when faced with the task of profiling the employment, income, education, housing and health status of the Aboriginal population in Kakadu National Park and acquiring some sense of how these might have changed over the previous 20 years, the Kakadu Region Social Impact Study had no readily available and comprehensive statistical information upon which to determine its case. This was due to a failure of the Australian Institute of Aboriginal Studies study to provide adequate quality data for the regional population, even as a baseline, let alone in the form of ongoing monitoring that was initially called for (Kesteven 1986).

Partly in response to such issues, and in the emerging context of corporate social responsibility, there has been a concerted effort in recent years by some major mining companies to address aspects of this regional information void with support for regional profiling in the East Kimberley (Taylor 2004a), the Pilbara (Taylor and Scambary 2005) and the West Kimberley (Taylor 2006b). Wider recognition of this requirement by Indigenous stakeholders is also indicated by land council and native title representative body support for such activity in the case of the Kakadu Social Impact Study (1997b), as well as in the regional studies listed above.

Such profiles are an important adjunct to the formulation and subsequent monitoring of company and Indigenous stakeholder actions designed to increase Indigenous participation in regional economies. Specifically, they help in three ways: by establishing the quantum of needs for regional planning (both present and future), by identifying particular opportunities and constraints for enhanced participation in regional economies, and by providing the demographic backdrop for assessment of the effectiveness of any actions undertaken, whether this is done using conventional or alternative approaches to the evaluation of outcomes from mining agreements (O'Faircheallaigh 2002b: 18–22).

Furthermore, with a focus on the demographic composition and population dynamics of mine hinterlands, regional profiles can assist in the identification of targets to meet particular objectives, particularly in providing an overall sense of the scale of potential undertakings. This adoption of targets for improving the situation of Indigenous peoples is an issue that has recently drawn the attention of the Aboriginal and Torres Strait Islander Social Justice Commissioner in discussions over the Australian government's reporting framework on Indigenous disadvantage (Calma 2005). Casting this in a human rights approach, the argument presented is for accountability in moving towards the achievement of identified goals within a defined time frame. In development discourse this invokes the principle of 'progressive realisation' and takes reporting to a new level by requiring that stakeholders make justification if there is no improvement in certain agreed benchmarks, or where there is improvement, to establish whether the progress achieved is at a sufficient rate (Calma 2005). While the levels and rates involved to assess such progress require regular calibration, fundamentally they require the development of a baseline from which to measure change.

Data on Indigenous populations in proximity to mining operations

The Minerals Council of Australia (MCA) estimates that 60 per cent of mining operations in Australia are 'neighbours' with Indigenous communities (MCA 2004: 5). To this extent, the prospect of broad-scale demographic analysis is invoked. However, any consolidated approach to the demography of Indigenous populations in proximity to mining operations would belie the variable availability, quality and applicability of population data, to say nothing of the diversity of social, economic and cultural conditions that exist across mining regions. The very existence of estimated different land use agreements referring to distinct population groups and regional configurations demands disaggregated analysis. However, as we shall see, some geographic scaling-up is also required because of the nature of data availability and the relatively small size of population groupings.

Population data

Two categories of population data are necessary to support the different roles required of regional demographic profiling. The first concerns data for monitoring as enabled by the provision of a statistical baseline of existing conditions against which an assessment of past and future change can be made. The second supports a predictive role, or at least an anticipation of the possible effects of proposed development set against future population scenarios. In constructing such profiles, a range of data are available from a variety of published and unpublished sources including the Census of Population and Housing and other Australian Bureau of Statistics (ABS) collections, administrative data sets held by Commonwealth and State government agencies, and statistical information from mining companies, regionally-based institutions, and Indigenous organisations. The array of such available data is substantial and some idea of the range, even for relatively small geographic areas, is provided in Table 3.1, based on experience in the Thamarrurr region of the Northern Territory (Taylor 2004a).

To date, regional profiles have involved the development of social indicators drawn from across the range of official data sets as shown in Table 3.1. For both Indigenous and non-Indigenous populations, these include demographic structure and residence patterns, labour force status, education and training, income, welfare, housing, justice and health status. Ideally, indicators for each of these categories should be established at the commencement of local mining operations and then subsequently in order to establish trends. To complete the profile, regional population projections should also be produced, typically for 20 years (roughly a generation). This may encourage forward thinking and an anticipation of needs—and the opportunity to respond to them before they are realised. Whilst acknowledging that the accuracy of projections diminishes with the length of projection period (Bell 1992; Smith and Sincich 1991), this capacity to project future population levels is an essential component of the preparation of regional profiles. All too often in Indigenous affairs, policy has been 'reactive', and by responding to historic levels of need thereby creating a constant sense of catch up. If mining agreements are to be effective catalysts for change what is required is a 'proactive' methodology which seeks to anticipate and plan for expected requirements—essentially a means of translating the content and intent of agreements into a required quantum of program and partner commitments over a given time frame.

Data quality

In establishing the relative social and economic circumstances of Indigenous and non-Indigenous Australians, there is considerable reliance on census data for many key indicators. This has a number of advantages given the comprehensive scope of coverage and the application of standard measures. However, there are drawbacks too. First of all, the five-yearly census means that available population data can be as much as seven years old when first publicly available, given the time taken to process census output. There is also the problem of coverage, both in terms of population counts and population characteristics, especially for the Indigenous population. To date, the net census undercount of Indigenous peoples has been estimated to be around 6–7 per cent, although this is likely to vary geographically, with much higher rates estimated for some remote regions (Martin and Taylor 1996; Taylor and Bell 2003). The application of a post-enumeration survey by the ABS in remote Indigenous communities for the first time in 2006 produced substantial estimates of net undercount of 24 per cent, 19 per cent and 12 per cent respectively in the Northern Territory, Western Australia and Queensland (ABS 2007). Non-response to census questions is also an issue (Taylor 1993), with relatively high rates of non-response observed for many Indigenous population characteristics. While little can be done about non-response for census characteristics, the ABS does establish post-censal estimates of the Indigenous population in an attempt to adjust for undercount and non-response to the Indigenous status question, and herein lies a solution to the problem of coverage raised above.

Table 3.1 Data items available for the Thamarrurr region from Commonwealth, Northern Territory and local government agencies

Population
ABS census counts and estimates of Indigenous and non-Indigenous population by 5 year age and sex for Wadeye town and outstations as a group.
Community census producing single year Indigenous and non-Indigenous age and sex capable of manipulation by community working groups into section of town, individual outstations, and clan groupings.
Clinic estimate of 'active client' Indigenous and non-Indigenous population by 5-year age and sex
Thamarrurr Housing Office population list of service population
Age and sex of Centrelink customers
Age and sex of regional residents on the electoral roll
Number of Indigenous persons registered with Medicare with a usual address in Thamarrurr

Labour force
Census data on labour force status, industry, occupation, hours worked
CDEP[b] participants by age, sex, and occupation
Centrelink data on Newstart and Youth Allowance payments
Local employer administrative records

Education and training
School enrolments by age, sex and grade level
School attendance by age, sex and grade level
School Multi Level Assessment Program (MAP) test results for Year 3 and 5 reading and numeracy
Enrolments: by training provider category by field of study by certificate level and accreditation category by outcome status by Indigenous status, age and sex
Census data on highest level of schooling achieved
Census data on post-school qualifications

Housing
Census estimates of housing occupancy rates
CHINS[a] and census data on housing stock by occupancy and number of bedrooms
CHINS data on housing stock by repairs needed
CHINS and census-based estimates of housing need

Health
Chronic disease incidence by age and sex
Growth characteristics of under-5s
Regional food costs compared to elsewhere in NT
Cost of family food basket
Fresh food variety, quality, availability
Unique hospital patients by Major Diagnostic Code (MDC), 5 year age and sex
Hospital patient separations by MDC by 5 year age and sex
Birth weights
Active client population for clinic by 5 year age and sex
Clinic staffing classification by Indigenous status

Justice
Reported regional property offences and offences against the person
Persons in adult correctional centres by last known address and birthplace
Juveniles in detention by last known address and birthplace
Adult conditional liberty caseload
Juvenile conditional liberty caseload
Conditional liberty order commencements

Welfare
Centrelink payments by type and $ amount (cells < 20 supressed)
Census estimates of employment and non-employment income

a. CHINS = Community Housing and Infrastructure Needs Survey.
b. CDEP = Community Development Employment Program.
Source: Taylor 2004b.

Basically, in situations of significant undercount, the census can be viewed as a very large sample survey with the key output being population rates rather than population levels (Siegel 2002: 495). Rates established net of non-response (on the assumption that the latter are evenly distributed for each population characteristic) can then be applied to population estimates—initially to the estimate for the census year, and then to any population projections from the census year on the assumption that the observed rates remain constant or change according to prescribed criteria. While an assumption of constancy might be seen as unrealistic, it should be noted that one of the unfortunate features of many Indigenous social indicators in mining regions over the past two decades (such as labour force status, income, education, and housing) has been their relative stability. It is also true that social indicator rates, by their very nature, are unlikely to drastically alter over short periods of time as they require substantial shift in levels in order to effect change (Hunter 1999). This is especially so among rapidly growing populations.

Whatever the approach to constructing regional profiles, it is crucial that they are based on reliable population estimates. Program-wise, this requires reliable breakdown into relevant policy age-groups: infants, mothers, school-age children, youth, young adults, middle-aged, and older people. Ideally, it also requires that statistical events in the population (such as employment numbers, school enrolments, hospital separations) are drawn from the same population universe, such that numerators are drawn from denominators in the calculation of rates. Unfortunately, under conditions of high inter-regional mobility and variable reporting of Indigenous status in administrative systems, this is not always the case (Cunningham 1998). What is clear, though, is that standard small area statistics as available from the ABS in the form of Indigenous Community Profiles provide only a starting point. Not only do these require ground-truthing in terms of cultural and geographic match, they are also limited in scope (and sometimes coverage), hence the need for additional data to be compiled from alternate sources.

Perhaps more telling from the point of view of data quality are concerns about the capacity of official data to provide a meaningful representation of the social and economic status of Indigenous people, especially in remote regions. Is it meaningful to measure one set of social, cultural and economic systems (Indigenous/local) using the tools, methods, and purposes of another (mainstream/national/global)? One view on this describes the process of census enumeration in remote Aboriginal communities as a 'collision of systems' and concludes that census questions lack cross-cultural fit and produce answers that are often close to nonsensical (Morphy 2002). Equally, while social indicators report on observable population characteristics, they reveal nothing about more behavioural population attributes such as individual and community priorities and aspirations for enhancing quality of life. Indeed the whole question of what the latter might mean and how it might be measured for Indigenous populations is only now being addressed by the United Nations Permanent Forum on Indigenous Issues in challenging the appropriateness of Millennium Development Goal indicators (United Nations 2006), an issue that has also been raised in the context of Australian frameworks for reporting disadvantage (Taylor 2008a). In working through these questions, we should be mindful that from an Indigenous perspective the very notion of measurement often carries with it the spectre of state control, and the implications of who is measuring what, for whom, and to what end should be crucial points for consideration, as demonstrated, for example, in respect of Maori (Smith 1999).

Thus, regional profiling using official statistical data can be seen simply as a method for 'rapid appraisal' and rightly criticised for lacking community input, thereby restricting its relevance and representativeness (Birckhead 1999; Walsh and Mitchell 2002). The danger is one of using data that have little relevance (to Indigenous stakeholders) and excluding those that do. Thus, along with the compilation of mainstream social indicators there is a need also to acquire information that reflects Indigenous priorities and understandings of what constitutes appropriate and sustainable development. This has some resonance with O'Faircheallaigh's (2002b) 'alternate' approach to evaluation which includes examination of the unfolding responses of Aboriginal actors to development initiatives. So far, however, only limited attempt has been made to incorporate such data into regional profiling (Taylor and Scambary 2005), although Scambary (this volume) has explored Indigenous development aspirations in considerable ethnographic detail and found these to be often at odds with more mainstream understandings of optimal outcomes from the interaction between miners and Indigenous people. Likewise, in a survey of traditional Aboriginal owners conducted in 2006 to establish what they wanted to do with their land, less than 13 per cent listed economic development as a first priority while more than one-third highlighted access, residence, land and sea management and cultural heritage (Balsamo and Calma 2007). Interestingly, these are all components of what O'Faircheallaigh (2006: 3–4) has referred to as the 'better life' that many Aboriginal people aspire to while living on land that they own. Partly for these reasons, the extent to which data of sufficient quantity and quality currently exist in Australia exist for the purposes of establishing meaningful baseline profiles remains a moot point.

Indigenous culture and measurement

A recent study of social and economic conditions in the Pilbara sheds some light on this issue (Taylor and Scambary 2005). This study sought the views of a selection of local Aboriginal people on conventional social indicators that were deemed to be representative of their parlous socioeconomic situation. One such comment reflecting on increasing levels of participation in mining employment was as follows:

> "Life is a bit better out here because of mining and those agreements. But my thing is my own kids, we're not pushing them into what we want them to be. It's up to them as individuals, I believe that's fair enough. They can go and work in the mine, but they will be men and will have kids of their own, and they need to be there for their own kids to learn and teach them their culture. Because it's about carrying on the traditional cultural ways teaching knowledge skills, and the country itself, all those kinds of things, the trees, the language, going to ceremony, going out on country. My kid's father is teaching our kids. His grandmothers and grandfathers, they passed on all the knowledge to him, making him understand who he is, he hasn't missed out on anything, he's got it all and he knows what his role is as a cultural man and in our cultural life. But some of those mining men aren't there for all that and that's no good." (Taylor and Scambary 2005: 58)

This statement has relevance for any discussion about the evaluation and monitoring of mining impacts. It highlights the fact that a positive mainstream measure of development (employment in mining) may have negative consequences for an Indigenous measure of well-being (carrying on traditional cultural ways). More to the point, it illustrates that a range of Indigenous views on the appropriateness of various indicators are likely to exist and that these may stand outside, and therefore be excluded from, mainstream indicator frameworks.

Not surprisingly, then, a consistent message to emerge from consultations conducted by the Australian Government with select Indigenous people and organisations regarding the measurement of disadvantage is the need to improve representations of Indigenous culture in formal reporting frameworks (Steering Committee for the Review of Government Service Provision 2005: 2.11). Although no explanation is provided as to what precisely is meant by the term 'culture' (Peterson 2005), a basic dilemma to emerge from these consultations is the difficulty of identifying single indicators given the diversity of Indigenous circumstances and societies across Australia. Furthermore, the fact that the construction of objective indices is more likely to be directed at informing government policy, and not necessarily Indigenous priorities and processes, means that the challenge is to satisfy the first requirement while at the same time producing measures that have widespread relevance to Indigenous peoples. Ultimately what is sought, then, is similar to the mechanism for illuminating the legal nature of native title for public discourse in Australia in terms of a 'recognition, or translation, space' that exists where Indigenous law and custom and Australian property law intersect (Mantziaris and Martin 2000). This conceptualisation of a legal 'recognition space' may be adapted to the area of social indicator development as illustrated in Fig. 3.1.

As inferred from the diagram, much of what constitutes important aspects of Indigenous world views, notions of productivity, appropriate structures of social relationships, land relationships, kinship rights and obligations, reciprocities and accountabilities (Altman 2005a; Martin 1995; Peterson 2005; Povinelli 1993; Schwab 1995; Trigger 2005)—in effect, different ways of life—is not necessarily brought to the level of public discourse (the intersect in Fig. 3.1), and is therefore not easily amenable to measurement. Even where measurement appears possible, distinct modes of Indigenous living and aspiration may be incommensurate with the broad goals of government policy to the point where they defy common understandings.

Fig. 3.1 The potential recognition space for indicators of Indigenous well-being

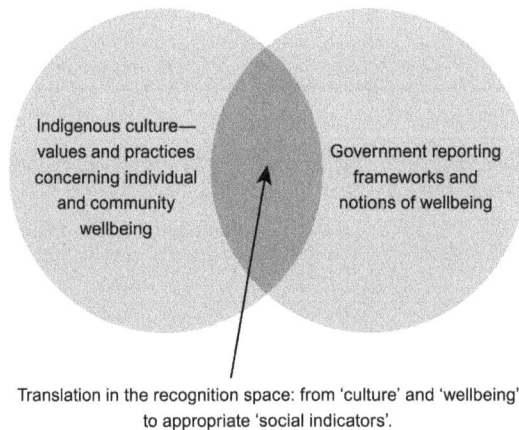

Indigenous culture— values and practices concerning individual and community wellbeing

Government reporting frameworks and notions of wellbeing

Translation in the recognition space: from 'culture' and 'wellbeing' to appropriate 'social indicators'.

Source: Taylor 2008c.

In the Australian context, for example, there is a clear contradiction between the desire of many Indigenous people to live in remote areas in small dispersed communities on traditional lands, and the general thrust of government policy intent on securing Indigenous participation in mainstream urban economies as a core means to enhance well-being. By the same token, elements of government reporting (certainly when it comes down to particular strategic change measures) may have little connection to Indigenous concerns and practices. An especially poignant example of this is provided by outputs from the Australian census which, because they are designed to represent the circumstances of mainstream

Australia, generate results for Indigenous peoples in remote settings that can have little meaning (Morphy 2002). As noted earlier, important elements of Indigenous customary economic activity, for example, can be overlooked entirely (Altman 2005a, 2007b; Altman, Buchanan and Biddle 2006), a problem noted for other Indigenous populations in developed country settings (Usher, Duhaime and Searles 2003).

The main focus of the diagram, then, is on the overlap where policy makers and Indigenous people can seek to build meaningful engagement and measurement. This is the area that allows for a necessarily reductionist translation of Indigenous people's own perceptions of their well-being into measurable indices sought by industry and governments. What is captured in this space is obviously far from the totality of Indigenous understandings of well-being, a point noted before in respect of Australian social survey data (Peterson 1996). As Peterson (2005) has also pointed out, without a common agreed view of different and shared perceptions of well-being, the danger is that indicators become ethnocentric and the notion that Indigenous people may have their own life projects is obscured by the pressing moral and political objective of achieving statistical equality that comes with policies of closing the gap and mainstreaming (see also McCausland 2005b; Tehan et al. 2006: 7-8). In working through these questions, the implications of who is measuring what, for whom, and to what end is therefore crucial. This cross-cultural encounter involves more than just recognition of difference—it requires the development of models of bi-cultural or partnership research involving negotiated design, methodologies and outcomes (Smith 1999: 173–8). According to the Aboriginal and Torres Strait Islander Social Justice Commissioner (Calma 2005), part of the means to this end lies in ensuring effective full participation of Indigenous people in all stages of data collection and analysis as an essential component of participatory development practice, much in the same way as has been achieved in New Zealand (Wereta and Bishop 2006).

In the sphere of economic activity alone, there is clearly scope within agreement-making processes to broaden the range of data within the area of overlap in Fig. 3.1 in a manner that better represents the priorities of Indigenous people. Indeed, it may be argued that agreement-making requires this (Tehan et al. 2006: 8). For example, if we return to the above quote from the Pilbara, then "carrying on traditional cultural ways" might be reflected in a range of economic activities associated with cultural and language maintenance, arts and crafts manufacture, land and sea natural resource management, and cultural tourism that are widespread across remote Australia (Armstrong, Morrison and Yu 2005; Hill et al. 2005; Toussaint et al. 2001). From a profiling perspective, the problem is that such activities are poorly quantified and often amorphously classified in labour statistics under the category 'Community Development Employment Program'.

To take the West Kimberley region as a case in point, there are numerous examples of significant impacts on local economic participation that derive from Indigenous priorities for economic development—the trochus hatchery at One Arm Point, the Pandanus Park freshwater crustacean project, the Manbana multi-species aquaculture hatchery and discovery centre in Broome, the Emama Nguda mud crab fishing enterprise in Derby, Goolarri Media Knowledge Centre in Broome—numerous small cultural tourism enterprises across the region provide commercial examples of this. Elsewhere, land and sea management activities include the Dugong and Marine Turtle project on the Dampier Peninsula, the Saltwater Country project along the north-west Kimberley coastline, the Coastal Landcare project at Broome, the Karrajarri Coastal Management at Port Smith involving Karrajarri Rangers, the Freshwater Sawfish project on the Fitzroy River, and the Rivercare project based on the development of a Fitzroy River Action Management Plan. Alongside these are activities aimed at developing consolidated 'looking after country' plans across the region combining the resources of the Kimberley Land Council, the Kimberley Aboriginal Law and Culture Centre, the Kimberley Aboriginal Pastoral Association, and the Kimberley Language Resource Centre. An example of this was the establishment by Kimberley Aboriginal Law and Culture Centre and the Yiriman Project of a pilot community ranger program at the Jarlmadangah Burru Community in 2006 with options now being explored to extend training and employment pathways for Aboriginal Rangers along the lower Fitzroy River. Soon to come on stream within the Ngurrara Native Title claim is the Great Sandy Desert—Warlu Jilajaa Jumu Indigenous Protected Area with proposals for further ranger programs. Altogether, 11 ranger programs currently exist across the Kimberley under the umbrella of the Kimberley Land Council and North Australia Indigenous Land and Sea Management Alliance with most of these heavily reliant on Community Development Employment Program (CDEP) funding.

Despite all this activity, there is no clear sense of overall impact for want of an appropriate instrument for measuring the collective number and nature of jobs created and the degree to which these might be further developed and supported to form part of a broad-based sustainable regional economy. In this regard, it is significant to note substantial movement on the part of the Commonwealth government with the establishment of the Working on Country program for Indigenous employment in activities such as fire management, feral weed and animal control and heritage site protection.

What hinterland? Defining the region

A fundamental issue for regional profiling, and one that is likely to assume variable character depending on the nature of different agreements, is the question of precisely which population (and therefore which geographic area) is implicated. Consideration of this matter has its origins in the original deliberations over the receipt of mining monies from the Ranger mine and the relevant provisions of the *Aboriginal Land Rights (Northern Territory) Act 1976* (Cwlth) (ALRA) that refer to 'areas affected'. While the political contestations in regard to this important matter are of interest in themselves, what concerns us here, with respect to profiling, is the outcome in relation to what ended up constituting the 'region' or geographic unit of analysis.

The Ranger case is instructive. Initially, 'the traditional owners of the Kakadu region' were to receive Ranger up-front monies (Levitus 1991: 156). While the 'Kakadu region' was not precisely defined, it was conceived as approximating Stage 1 of Kakadu National Park and the Ranger Inquiry land claim identified 107 people as the traditional owners of that area (Levitus 1991: 157). However, in creating an Association to hold dealings with traditional owners, the Northern Land Council adopted a more inclusive approach to compiling Association membership. As a consequence, by 1979, membership of the new Gagudju Association was much larger and comprised individuals, 'connected with the Ranger country, either through blood ties, intermarriage, clan relationships or some shared dreamings' (Levitus 1991: 157). By the mid-1990s, however, the Mirrarr Gunjeihmi clan challenged the validity of the Gagudju Association to receive 'areas affected' monies and legal opinion supported their view that the ALRA reference to area affected referred to the physical area of the Ranger mining lease rather to the spatially much wider notion of social affectedness (Altman 2007b, this volume Chapter 2).

All of this greatly complicates matters in terms of circumscribing a definitive geographic area for the purposes of regional profiling. It also shows that where there is intention to construct social indicators from secondary sources for a population defined on the basis of cultural criteria then this can be rendered difficult by the sort of ambiguities illustrated above. This aside, the fact that most official and administrative data are available at fixed aggregate levels only and mostly for relatively large areas (certainly beyond a mining lease area) results in a degree of inflexibility. In the Kakadu Social Impact Study, this area turned out to be that bounded by Stages 1 and 2 of Kakadu National Park as these matched ABS geographic boundaries (Taylor 1999). For the most part, then, the geographic areas for which statistical information is available (including administrative data) are dictated by the boundaries set out in the ABS' Australian Standard Geographic Classification (ABS 2006a) and the Australian Indigenous Geographic Classification (ABS 2002).

Accordingly, in profiling socioeconomic circumstances, the tendency has been to take a broad regional, rather than limited local, perspective on the geography of mine hinterlands. This is not just because of the sorts of issues raised above, but also because a key interest of some mining companies and Indigenous stakeholders is to build regional economies beyond the mine gate (Barker 2006;

Harvey 2002a). One manifestation of this is the negotiation of Indigenous Land Use Agreements (such as the Argyle Participation Agreement) that can have effect over substantial geographic areas. In the lead up to this agreement, discussions with the Kimberley Land Council about what constituted the region of interest for baseline profiling focused initially on the area covered by the Good Neighbour Program involving communities closest to the Argyle mine site, but then this widened out to incorporate the whole of the east Kimberley excluding the southern part of Halls Creek Shire, but including Kalumburu (Taylor 2004a). Likewise, the *Century Zinc Project Act 1997*, enacted in relation to the Gulf Communities Agreement, referred to a large area known as the Carpentaria-Mt Isa mineral province. Such interpretations necessarily expand the scope of any analysis beyond the potentially narrow geographic bounds of immediate mine hinterlands to encompass more functional definitions of 'area affected' based on some measure of regionally integrated social, economic and administrative interactions. In the Pilbara, for example, the physical separation of most Aboriginal people from actual mine sites means that 'local' labour is likely to be drawn from across the whole of the ABS' Pilbara Statistical Division which more or less represents the jurisdiction of the Pilbara Native Title Service under the umbrella of the Yamatji Marlpa Barna Baba Maaja Aboriginal Corporation (Taylor and Scambary 2005).

Although regional perspectives clearly emerge in practice, what is lacking is any operational definition or criteria for guidance on regional selection, except to say that there seems to be some trade-off between regions identified according to customs and traditions versus the need for economies of scale and recognition of existing service delivery frameworks in development planning. This is captured by the Northern Territory Government's Stronger Regions Policy that initially envisaged negotiated regions based on the following criteria:

> an area that the people in it see as a region and that the government agrees should be treated as such; where a reasonable community of interest exists; where there is capacity to achieve economies of scale in the achievement of outcomes; and where there is demonstrated capacity or need for whole of community action to cooperate in the achievement of shared objectives (Northern Territory Government 2003).

While such groupings might appear intuitively sound, it should be noted that complexities are almost certain to arise in seeking to establish boundaries for the purposes of representing regional 'communities of interest' with 'shared objectives'. Useful insights into the nature of such complexities and how they might impact on attempts at regional planning are available from Sutton's (1995) critique of Davis and Prescott's (1992) work on Aboriginal boundaries, and Morphy's (1999: 36) critique of the Reeves (1998) proposals for reform of the ALRA.

Indigenous population trends in Australia

For the most part, Australian mineral provinces occupy large swathes of country across the remoter parts of the continent. In such areas, significant demographic shifts have been underway since colonisation although the defining feature of regional demography remains the high Indigenous share of population and the relatively high Indigenous rate of natural increase (Taylor, Brown and Bell 2006). By way of background, it is worth noting that parallels exist between the modern-day classification of remote areas and the historic distinctions drawn between 'colonial' and 'settled' Australia in recognition of the much higher proportions of Indigenous people in remote areas, and the somewhat different manner of their incorporation into wider social and economic structures (Rowley 1971).

This spatial framework also provides essential context for understanding the substantial transfer of land back to Aboriginal ownership across remote regions that has occurred in recent times, with the prospect of more to come via land purchase and successful native title claims (Altman, Buchanan and Larson 2007; Pollack 2001). One prominent long-term analyst of demographic trends in remote areas views this land transfer as an important element of the post-productivist transition in Australia's rangelands with newly recognised land values often lying outside the old economy, being more culturally-based (Holmes 2002). These values are manifest in the emergence of a distinct settlement structure on Aboriginal lands involving the formation of numerous, dispersed, small, Indigenous communities, especially in the Northern Territory, Western Australia and the far north of South Australia and Queensland. Most, if not all, of these communities required no modern economic base, nor have they subsequently acquired one, at least not in a manner beyond the combined provisions of a sizeable state sector, a limited private sector, and a customary sector of variable size. The term 'hybrid economy' has been coined to describe this structural arrangement (Altman 2005a). Across remote areas, a total of 1112 such communities were identified by the Community Housing and Infrastructure Needs Survey (CHINS) in 2006 with a total estimated usual population of 80,500 (ABS 2006b: 17). The vast majority (838, or 75 per cent) of these were very small in size with less than 50 persons, although collectively these very small places accounted for only 10,200 persons, or 13 per cent of the total in remote discrete communities (ABS 2006b: 5).

For reasons of differential population dynamics, the Indigenous population in remote parts of Australia grew by 23 per cent between 1981 and 2001 while overall non-Indigenous population growth over the same areas has been negative since 1986 (Taylor 2003). Away from the larger mining towns and service centres of remote Australia, Indigenous peoples are increasingly the majority. These trends are leading to a rising Indigenous share of remote area population and there is every reason to expect that this will continue. While there is no denying that minerals development has been a major stimulus to population growth in these areas since the 1960s, it is also true that this has been highly variable due to boom and bust in commodity prices, while part of the ongoing dynamic has been a distortion of underlying non-Indigenous population growth rates due to the expansion of fly-in-fly-out arrangements (Storey 2001).

Across the vast arid zone, the Indigenous population is projected to rise from 39 800 in 2006 to 47 000 in 2021 while the non-Indigenous population is projected to decline from 135 000 to 124 000 (Taylor Brown and Bell 2006) producing an increase in the Indigenous share of population in Arid Australia from 23 per cent to 27 per cent. In the combined regions across the wet tropics from Cape York to the Kimberley, similar projections to 2016 indicate a rise in Indigenous population from 25 600 to 32 400 representing an increase in population share from 38 per cent to 42 per cent (Taylor and Bell 2001). Even in the more economically dynamic remote regions of the outback, such as the Pilbara, the Indigenous share of the usually resident regional population is projected to continue to rise from 16 per cent in 2001 to 18 per cent by 2016 (Taylor and Scambary 2005). Some indication of the sorts of figures involved for Indigenous growth in selected mining regions is shown in Fig. 3.2 using 1996 Census based cohort-component projections, although ideally such projections would also be informed by economic impact forecasting (Phibbs 1989).

Employment implications of Indigenous population growth

Given that commitments to Indigenous employment and training form a key part of many mining agreements, one of the key implications of this high Indigenous population growth is the manner in which it interacts with labour demand and supply. In recent years, the thrust of both government and minerals sector policy has been towards increasing Indigenous employment in mainstream jobs. If we take an historic view of this we can see that while the numbers of Indigenous people employed in mining regions may have increased over the past 30 years or so, the rate has not because of a failure of jobs growth to keep pace with population growth. What then is the scale of the task ahead if the aim is to increase the rate of Indigenous employment? To establish this, we can use the projection of the future size of the working age population and consider this against variable scenarios in terms of regional employment rates.

Fig. 3.2 Indigenous population projections 1996–2016 in selected mining regions

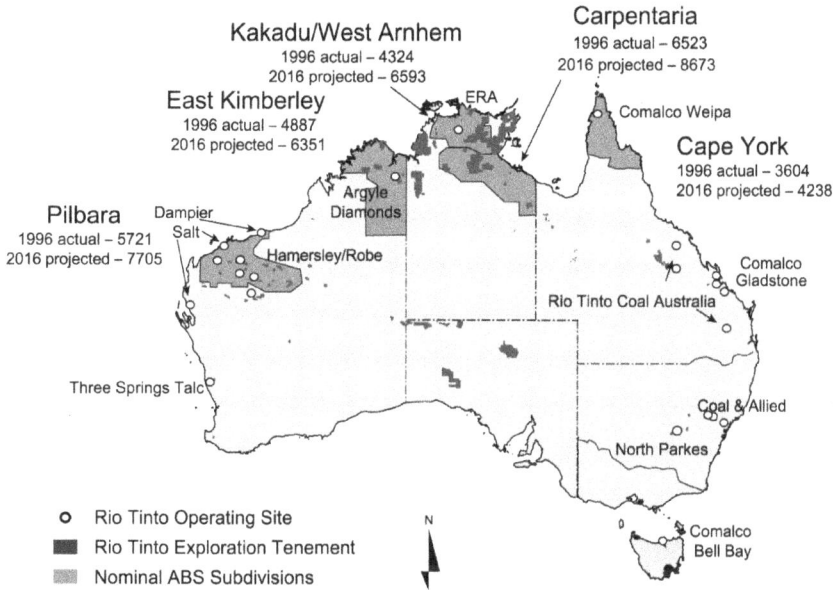

Kakadu/West Arnhem
1996 actual – 4324
2016 projected – 6593

Carpentaria
1996 actual – 6523
2016 projected – 8673

East Kimberley
1996 actual – 4887
2016 projected – 6351

ERA

Comalco Weipa

Cape York
1996 actual – 3604
2016 projected – 4238

Pilbara
1996 actual – 5721
2016 projected – 7705

Dampier
Salt

Argyle
Diamonds

Hamersley/Robe

Comalco
Gladstone

Rio Tinto Coal Australia

Three Springs Talc

Coal & Allied

North Parkes

o Rio Tinto Operating Site
■ Rio Tinto Exploration Tenement
▨ Nominal ABS Subdivisions

N

Comalco
Bell Bay

Source: Harvey and Brereton 2005.

The West Kimberley region provides a useful example. By 2021, the Indigenous adult population of the West Kimberley is projected to increase by 36 per cent (or an average of 2.1 per cent per annum) to reach a population of 9156—an increase of almost 2500 persons. What are the implications of this growth for future jobs needs if certain employment rates are to be achieved? Two scenarios are explored in Table 3.2. The first considers the number of Indigenous jobs that would be required in the West Kimberley by 2021 if the 2006 Indigenous employment/population ratio were to remain unchanged at 53.5 per cent (inclusive of CDEP). The answer is 4 898 jobs, or an additional 2 381. The second scenario considers the future job requirements necessary to raise the Indigenous employment/population ratio to that recorded for non-Indigenous residents (79.3 per cent). This requires the number in work to reach 7261—massively in excess of the current level. In effect, to close the regional employment gap with the non-Indigenous workforce, the number of Indigenous people in work across the West Kimberley would need to almost triple over the next 15 year period, with an additional 316 jobs created and then occupied each year.

Table 3.2 Extra Indigenous jobs required in the West Kimberley by 2021

Employment/population ratios in 2006	Base employment 2006[c]	Total jobs required by 2021[c]	Extra jobs required by 2021
53.5[a]	2517	4898	2381
79.3[b]	2517	7261	4744

a. The Indigenous census-derived employment/population ratio in 2006 inclusive of CDEP.
b. The non-Indigenous census-derived employment/population ratio in 2006.
c. Based on projections of Indigenous adult population by 2021 (9156).
Source: Taylor 2006b.

Implications for government and industry

As noted at the outset, the role of applied demography is to provide an essential quantum to discussions of needs, aspirations, and policy development. The basic message conveyed from such applications to date in mining regions across northern Australia is that little had been achieved over the decades since mining commenced in terms of enhancing overall Indigenous socioeconomic status, and that this is now exacerbated by rapid population growth. Despite current unprecedented demand for Indigenous labour in the formal economy, the capacity of labour to respond to this remains substantially constrained by limited human capital. As a consequence, many in the Indigenous population will continue to experience structural dis-engagement from mainstream work in the absence of substantially enhanced infrastructure and services to redress historic exclusion.

Of course, in pursuit of a social licence to operate, major corporates are active in engaging Indigenous workers with many mine sites adopting target quotas. But such is the depth of supply-side disadvantage that a major challenge lies ahead in meeting these targets (certainly in a collective sense) given that they are likely to come close to exhausting entire local supplies of employable labour. For this reason, companies are increasingly investing in remedial programs to enhance work readiness and to address structural barriers in meeting 'fitness for work' requirements (Tiplady and Barclay 2007). Even so, if current targets were to be achieved, the additional jobs created might only suffice to keep pace with the growth in the Indigenous working-age population. Thus, while much might be accomplished by the mining sector in the years ahead in terms of raising levels of Indigenous employment, little change might be discernable in terms of aggregate social indicators, with a large component of the population remaining marginalised. To avoid this, it is essential that all sectors of regional economies should be utilised for Indigenous engagement, including activities associated with Indigenous priorities and aspirations in arts and craft manufacture, land and sea resource management, and cultural tourism. However, whatever development options are pursued, substantial constraints on participation remain to be overcome across the spectrum of social and economic conditions.

To indicate the scale of some of these constraints that have been quantified, Table 3.3 provides indicative proxies of labour force exclusion in the Pilbara for 2006. It is necessary to bear in mind that the adult population for that year was projected to be 4,759. What this shows is that the vast majority of Indigenous adults in the region do not have full schooling; or a qualification; around half of adults remain outside the labour force; many are hospitalised at any one time; others are subject to chronic conditions requiring strict management regimes; many are arrested and incarcerated (especially young males); and feeding into this adult realm are relatively low achievers from the education system. In any event, the potential for prolonged and productive workforce participation on the part of young people is severely curtailed by premature mortality.

Table 3.3 Summary indicative proxies of Indigenous labour force exclusion, Pilbara region, 2006

Population aged 15 +	4759
Has no post-school qualification	4200
Has less than Year 10 schooling	1500
Not in the labour force	2190
Hospitalised each year (all persons)	2800
Has diabetes (25 years and over)	1020
Has a disability	1020
Arrested each year	1050
In custody/supervision at any one time	310
Achieves Year 7 literacy (current attendees)	60%
15 year olds surviving to age 65	<50%

Source: Taylor and Scambary 2005.

From a policy perspective, levels of economic exclusion on the scale implied here raise questions about the adequacy of government resourcing to meet the backlog of disadvantage that has so obviously accumulated in many mining regions. Looking ahead, this raises questions about the costs to industry, to government, and to Indigenous people, if social and economic conditions remain the same as currently experienced. Basically, these costs will simply escalate in line with the growth in population. However, to properly assess adequacy in this context, it is not sufficient to consider amounts expended by governments on Indigenous programs separately from the key questions of whether such amounts are commensurate with the scale of the task of overcoming disadvantage, and whether they are equitable on a per capita basis when compared with spending in the States and Territories and nationally on people in similar circumstances.

It is interesting to note that the only analysis to have measured spending in this way for a single Aboriginal community (Wadeye) in the Northern Territory uncovered gross inadequacies and inequities in crucial areas of capacity building such as education, training and infrastructure with associated high expenditures in health, welfare and incarceration (Taylor and Stanley 2005). If similar inadequacies were to exist in communities across mining regions, and at this stage this is not known, then the level of government intervention aimed at overcoming disadvantage among the growing Indigenous populations of these regions would not only be found wanting, it would simply be 'funding into a deficit' (Ah Kit 2004). This reflects the regressive nature of the link between demography and economy in contemporary Indigenous Australia and it means that governments, and industry for that matter, can either invest up-front to build capabilities, or pay heavily in the future to manage the social and economic consequences. Whatever the case, a fiscal response is unavoidable.

Conclusion

With approximately 18 per cent of the Australian continent under some form of Indigenous tenure (Altman, Buchanan and Larson 2007), and with this set to rise via native title determinations and land purchases, the demand for statistical information on Indigenous groups as proprietors of territory is also growing in the context of formal agreement-making. As with the broader government agendas of closing the gaps, applied demography has found a natural and successful disciplinary niche here by exploiting the rich seams of census, survey and administrative data that make up the burgeoning Indigenous statistical archive, even at local and regional levels. In particular, it is the predictive capacity of the demographic repertoire that has mostly caught the eye of regional stakeholders (Harvey and Brereton 2005) with its fiscal opportunity-cost message that business as usual is simply not an option in Indigenous affairs because of the weight of population momentum. However, there are two important constraints on the efficacy of this contribution that need attention if the interests of Indigenous stakeholders are to be truly represented and if the scope of applied demography is to be extended.

The first of these relates to geography and concerns the long-standing conundrum of determining which area/peoples are affected/implicated by mining. In regard to this, it remains the case that analysts are constrained by the configuration of official ABS and other administrative boundaries that are used for the collection and dissemination of official statistical data. While it is true that the Australian Indigenous Geographic Classification of the ABS attempts to best reflect the spatial distribution of the Indigenous population—to enable a 'demography in situ' as it were (to use Kreager's (1982) term describing demography built from the ground upwards)—in some ways this misses the point entirely. The primary organising principles of Indigenous social formation are both spatial *and socio-relational* (Morphy 2007) and these invariably do not coincide to produce discretely bounded social groupings that neatly mesh with units of the Australian Indigenous Geographic Classification. In a nutshell, the statistical geography available to analysts is unlikely to provide a demography of Indigenous polities with rights and interests in particular places, or agreements.

In many ways this highlights an important distinction raised by Rowse (2008) between Indigenous populations and Indigenous peoples. Our current demography refers to Indigenous populations revealed by standard identification questions and is best suited to the provision of citizen rights. What it does not necessarily provide for are the interests of groups of Indigenous people in proprietory rights, in particular over areas of land. All across Australia we are witnessing a growing discrepancy between the best-intentioned of statistical output frameworks and the actual needs of Indigenous land-holding groups for an ethnographically-informed demography suited to their aspirations for managing the Indigenous estate via land use agreements.

The second constraint refers to the nature of available data themselves. While information on vital demographic events are constrained only by the degree to which this is gathered in respect of any population, the greater concern is to do with information on population characteristics as well as broader epistemological questions surrounding notions of well-being and what these imply about appropriate variables to be measuring. As things stand, the social, economic, and even some cultural, features of Indigenous populations are invariably established via mainstream categories. Whether these coincide with Indigenous categories in any given situation, and whether such categories can be identified, measured and are desired, remains a moot point.

4. Aboriginal organisations and development: The structural context

Robert Levitus

In recent decades, the debate over mining development and Indigenous peoples has broadened, both geographically and thematically. The resistance paradigm that asserts the rights of communities against a new industrialised wave of dispossession has been sustained and applied to a succession of new case studies (Cultural Survival 2001; Downing et al. 2003; Hyndman 1994; Lane and Chase 1996; Roberts 1978). Alongside that, and especially in developed countries such as Australia and Canada, attention has shifted to work within a paradigm of engagement, negotiation and management (Howitt 2001: 208–65; Render 2005; Vachon and Toyne 1983). Much discussion now revolves around the achievement of best-practice processes to maximise the short and long-term benefits that mining can offer, and mitigate or compensate for the problems it causes (Hill 1999; Indigenous Support Services and ACIL Consulting (ISS/ACIL) 2001; O'Fairchcallaigh 1996, 2003b; Sosa and Keenan 2001).

The most widely recognised benefit is the direct injection of earned income into the Indigenous domain, either by employment of Indigenous people in the mining operation or support for Indigenous businesses created to service the project (Cousins and Nieuwenhuysen 1984). A problem affecting both these economic inflows is their long-term sustainability, especially in the local area. Once the mining operation ceases, Aboriginal employees with skills and experience may be faced with a choice of returning to an environment of unemployment or underemployment, or moving to areas with an active job market. Similarly, businesses may have to fold or seek custom elsewhere.

A more indirect way in which mining companies have sought to transfer benefits to Indigenous communities is through the funding of local organisations which apply the received resources to various programs of their own design. In some cases these organisations' revenues will come in part from their role as business operators and thus face the same risk of mine closure. More importantly, however, mining agreements can direct substantial revenue flows, in the form of entitlements to an agreed percentage of mine income, to local organisations, allowing or even requiring them to accumulate capital funds over which they exercise long-term trusteeship. This is an important but under-utilised way in which mining companies can contribute to the preconditions for long-term remote-area Indigenous development that is not ancillary to the mining operation itself.

Aboriginal organisations funded from mining revenue sit at the intersection of two policy discourses in Australia. The first is that adverted to already: that carried on within and around the mining industry itself about benefit packages for communities affected by mining operations. While also part of an ongoing national and international debate over social justice towards Indigenous people, this concern is internally motivated by increasing corporate recognition of the business case for preserving a social license to operate in remote areas where the population is often predominantly Indigenous. The second is the ongoing discourse around national policy in the area of Indigenous affairs. Within the paradigm of self-determination that had currency at the Federal level from the 1970s to the 1990s, Aboriginal organisations were strategically central. During those years also, the issue of the social license to operate came to the attention of some mining companies operating in Australia, so that the emerging industry attitude began to align with the national policy attitude. State government recalcitrance, especially in Queensland and Western Australia, and industry reaction, especially from the peak Australian Mining Industry Council (now Minerals Council of Australia), meant that this was never a uniform trend on either side. Moreover, the most recent phase of Federal Aboriginal policy-making[1] has substantially abandoned the self-determination paradigm. Nevertheless the recognition of a form of native title in Australian law and a growing trend towards agreement-making have developed to sustain and expand the niches occupied by local Indigenous organisations, and the industry's preparedness to acknowledge their standing.

The purpose of this chapter is to consider Aboriginal organisations in their intended role as agents of development. It aims to identify a set of structural conditions under which they operate and which bear upon their capacity to implement development. My focus here thus falls within the paradigm of 'capacity building' that has attracted recent attention (Commonwealth of Australia 2004: 111–65). It is distinct from the common emphasis upon operational deficits in matters such as skills, knowledge and leadership suffered by many Indigenous organisations, but it does relate, especially in one of the later sections, to questions of governance. While the primary concern, consistent with the theme of this volume, is those organisations funded by mining revenue, they make up a small sub-set of Aboriginal organisations in Australia, and share the same general policy origins as the others. Thus, much of the following discussion has a wider focus. However, while the structural conditions that I identify are not unique to mining-funded organisations, I argue that several factors, including recent changes in the Indigenous affairs policy environment, cause these conditions to impinge upon such organisations in particular ways. Depending

---

[1] This is being written immediately following the election of November 2007 in which the conservative Liberal-National Party coalition was replaced by a Labor Government.

on what share of their total resources are accounted for by mining, this has implications for their capacity to define a developmental path on behalf of their constituents.

After looking in the next section at the emergence of Aboriginal organisations in policy history, I discuss four structural conditions affecting their capacity to effect developmental change on behalf of an Aboriginal constituency. Central to this analysis is Charles Rowley's conceptualisation of Aboriginal organisations as a carapace. As such, they are simultaneously a transactional boundary and a point of articulation between external agencies and an Aboriginal domain. The structural conditions that I identify bear principally on these qualities. The first is the level of political authority at which the organisation articulates with its resource providers in the wider society, usually a state agency, sometimes a mining company. The second is the manner in which resources are transferred across that point of articulation into the jurisdiction of the Aboriginal organisation. The third is the availability of alternative sources of servicing and supply, independent of the organisation, to which its members or clientele may have resort. The fourth is the endogenous relationship between the organisation and the Aboriginal domain that it exists to service and/or represent. I then consider the particular position of organisations receiving funding from mining agreements with respect to these structural conditions. In the penultimate section, I discuss a central developmental function incumbent upon organisations located at this resourcing interface.

The Aboriginal organisation in policy: carapace and domain

The idea of corporatising the Aboriginal interest emerged from the progressive political thinking of the 1960s. Those corporate entities became a critical instrument through which the new policy paradigm of self-determination was to operate. Rowley, and like thinkers such as Strehlow (n.d.: 19–24) and Coombs (Rowse 2000a), argued for respecting the integrative values of the Aboriginal social group, and allowing room for Aboriginal group choice in implementing social change. They saw the Aboriginal domain as a basis for new forms of planning and action, as 'an opportunity for some real political science, ... for some real government' (Rowley 1966: 236). Rowley and Coombs wanted to admit the Aboriginal group as an agent of development, given an expanded conception of what development might mean. So the design of suitable forms of Aboriginal organisation and the fostering of an Indigenous leadership were strategically central to these policy reformers (Rowse 2000a: 31–3).

This was a decisive step in the field of Aboriginal development. Prior to this, Aborigines figured as welfare recipients or units of labour. Even where congregated as 'communities' on reserves, settlements and missions, Aborigines participated in the developmental programs of the assimilation era mainly as workers in internal servicing and primary production jobs, making sure that the structures of dependency were at least functional. The creation in the 1970s of Aboriginal councils and servicing agencies, and their recognition as an entry point for government moneys payable to those settlements, were intended to promote the conditions for self-determined development, that is, the endogenous determination of the future of those places as Aboriginal.

Self-determination thinking was open to the spread of such arrangements throughout Aboriginal Australia for almost any purpose relevant to local needs (Anderson 2004: 260; Commonwealth of Australia 2004: 111–12; Rowley 1972: 439–41). They were adopted even where Aborigines lived among whites, and the physical separateness characteristic of remote communities was absent. The adoption of the legally incorporated form across all areas of Aboriginal interest proceeded apace, so that from the mid-1970s many thousands of associations, services, centres, councils and corporations were registered under either the Commonwealth's *Aboriginal Councils and Associations Act 1976*,[2] or corresponding State and Territory legislation. Rowse has written about this explosion of Aboriginal organisations across Australia as a central manifestation of the shift from assimilation to self-determination (Rowse 2000b: 1515–16; 2005: 214–19).

That policy step has consequently had major ramifications in the structuring of resource flows and political activity in all Aboriginal domains. The 'community', given legal personality by organisation into corporate forms, was deemed to be the initiator of development. From the perspective of the state, Aboriginal organisations were each a carapace beneath which the business of development could be allowed to proceed. These Indigenous legal entities thus carried the weight of the state's expectations that it could fund self-determining Aborigines to deliver development (see Batty 2005).

The two terms, domain and carapace, require comment. In his discussion of the concept of the Aboriginal domain, Rowse (1992: 19–21) substantiates it primarily in terms of content: Aboriginal people, using Aboriginal languages, organising themselves according to Aboriginal values for the pursuit of Aboriginal priorities. He also adverts to a criterion of size, but does not insist upon it, noting that some writers have applied the term to Aboriginal fields much smaller than, for example, the west Arnhem regional domain described by von Sturmer. Indeed Rowse distinguishes domain from 'enclave', a term he prefers to apply to such larger regions. The large-scale mining projects considered in this volume occur in regions where many Aboriginal people live in discrete settlements that would satisfy such a macro conception of the Aboriginal domain. Others, however, live in towns among white neighbours. Like Rowse, therefore, I do not wish to impose any requirement of size. Any arena in which affairs are consistently conducted

---

[2] Recently replaced by the *Corporations (Aboriginal and Torres Strait Islander) Act 2006*.

in an Aboriginal idiom and informed by Aboriginal meanings, such that Aboriginal people expect such a mode of behaviour to occur there, is an instance of the Aboriginal domain.

In the last two decades, the concept of domain has been overtaken by new modes of conceptualising relations between the Indigenous and non-Indigenous in Australia (Hinkson and Smith 2005). I retain the concept here because it was central to the model of incorporation by which Rowley and his colleagues thought self-determination could be activated. Domain and carapace were the two structural halves of that model. In this chapter, inquiring into the activation of that model in the context of development, I want to ask how those constituent concepts fare in the face of what I argue to be the structural preconditions of such a role. In other words, we can judge the adaptability and durability of Rowley's conception by seeing how well it has meshed with those preconditions. Indeed, in such an analytical context, the concept of domain remains, in my view, a necessary tool for considering relations of articulation between collectivities. Moreover, the ethnographically minimal conception of domain being proposed here implies none of the hermetic or endogenously self-reproducing characteristics that have been the main cause for criticism of the concept.

Rowley (1972: 423, 429) conceived of the Aboriginal organisation as a carapace in the sense of a protective cover for the localised Aboriginal domain. My use here of the concept of carapace does not depend on the relationship between organisation and domain taking any particular ethnographic form, but rather arises from the positioning (and self-positioning) of organisations within policy. Later discussion will refigure the carapace as often more partial and fragmented than allowed by the initial vision of policy planners and intellectuals.

Viewed from the outside, it is clear that limitations upon the capacity of Aboriginal organisations to facilitate the self-determined development expected of them are implicit in the niche that has been created for them in public Aboriginal affairs. Historically, colonialism forged relations of articulation that now make Aboriginal society everywhere a part-society. To borrow a metaphor once prominent in anthropological theory, responsibility for the social reproduction of Aboriginal society is therefore nowhere located entirely within an Aboriginal social universe. It always draws to some extent on external sources of supply that are under non-Aboriginal control. This is so even in those remote areas where Indigenous domains are more socially self-contained. As discussed above, Aboriginal organisations are themselves an artefact of a particular moment in the policy management of those relations of articulation. They occupy niches created when the state decided that the points of articulation between Australian society and the various manifestations of Aboriginal part-societies that it encapsulates, should be made more permeable to Aboriginal participation and receptive to Aboriginal concerns. There they proliferated to a collective position of prominence.

As carapace, the Aboriginal organisation is an institution that creates a formal transactional boundary for certain purposes between its membership or clientele and the outside world. It is at the same time a conduit for external resources flowing into the Aboriginal domain. In the discussion that follows I want to look at that articulatory role and what it implies for Aboriginal development. The institutional boundary just referred to was seen to be essential if the development enabled by incoming resources was to be self-determined. The policy function of the organisational carapace as a point of articulation between Aboriginal and non-Aboriginal domains is thus central for realising self-determined development. There are at least two other structural conditions, however, that affect its potency in that respect. The first is the political level at which the point of articulation is established, and the second is the mode of transfer of resources from the non-Aboriginal domain.

Condition 1: Political level of articulation

The design of the carapace—the geographical area, functions, Aboriginal population and other organisations that it will encompass—has in past years been a major site of public debate over Aboriginal political futures. It seems that Rowley's (1972: 412–13, 424) initial conception presumed an organic correspondence between the organisation and the local Aboriginal group, and the Council for Aboriginal Affairs in the early 1970s similarly thought in terms of 'a politics of localised empowerments' (Rowse 2000a: 107–8). In time, political evolution and the legislative recognition of new rights at State, Territory and national levels produced correspondingly higher levels of Aboriginal organisation that typically discharged servicing and distributional responsibilities internally and lobbying and representative functions externally. Land councils and the Aboriginal and Torres Strait Islander Commission (ATSIC) have been the most familiar of these.

A domain of Indigenous control in the management of some resources intended for certain functions, is created beneath each such carapace. To that extent, the relationship between domain and organisation is reciprocal, such that the disposition of resources through the organisation tends to re-figure Aboriginal political relationships in reference to it (Martin and Finlayson 1996: 6–7; Sutton 1998: 69). Where the carapace exists at a high level of political aggregation, regional or larger, and other more localised organisations relate to that higher level as clients or constituents, that initial point of articulation with non-Aboriginal funding providers overlooks a geographically broad and internally differentiated Aboriginal domain of decision-making and service provision.

The structures operating in the Northern Territory under the *Aboriginal Land Rights (Northern Territory) Act 1976* (Cwlth) (ALRA) for distribution of moneys from resource developments on Aboriginal land, or those proposed by Pearson (2000a: 67–73) for a new model of governance in Cape York, are such instances. Smith (2002: 25–7) has proposed a model of jurisdictional devolution that she calls 'regionally-dispersed layered community governance', whereby major communities are conceived as bases of Indigenous authority and responsibility. They would then act 'as the starting point for aggregation at the sub-community and regional levels', servicing smaller satellite settlements, and being themselves aggregated for particular functions that are more effectively discharged at higher political levels. It seems clear, however, that any institutional arrangement established at a supra-community level would require its own funding entitlements from the state for stability and effectiveness, lest it fall away with each successive community secession. Thus the structural depth and geographical breadth of the Aboriginal domain depends in part on the vertical positioning of that linkage with the non-Aboriginal world that is made at the point of entry of external financial transfers.

Condition 2: Mode of resource transfer

The other important set of mechanisms affecting the jurisdictional power of the Aboriginal carapace relate to the manner of payment and extent of oversight of funds paid into the organisations concerned. In part this is the familiar issue of accountability. The application of external public funds to internal Indigenous programs places Aboriginal organisations in a mediating role between external accountability requirements and internal cultural conventions and proprieties over matters of access, distribution and control (Commonwealth of Australia 2004: 128–34; Macdonald 2004: 62–4; Martin and Finlayson 1996: 1–8). Much comment has recognised the difficulties that office-bearers and staff face in attempting to effect positive change in the face of multiple expectations that are both incompatible amongst themselves and unsympathetic to the overall operating environment (Austin-Broos 2003: 128; Sullivan 1996: 95–7; Thorburn 2006). The associated stresses may perpetuate the managerial dominance of relative outsiders who have nothing to lose but the job itself.

The complexity of both applying and accounting for external funds, and the weight this places on small organisations with limited skills, has been another difficulty. Such demands often leave no time for strategic planning, much less for the 'vision thing'. Pearson (1999: 33) criticises the 'hydra' character of government articulation with Cape York communities, of multiple parties and programs, and demands a more coordinated and holistic regional approach to resource use in which government plays a junior role (and see Gerritsen 1982: 18–19; Macdonald 2004: 44–5; Smith 2002: 6, but cf. Gray and Sanders 2006: 22–3, 25; Sanders 1993). A recent phase of policy innovation in Aboriginal affairs has emphasised coordination of agencies' programs across all functional areas through the Office of Indigenous Policy Coordination and a number of Indigenous Coordination Centres at the national and regional levels respectively (Office of Indigenous Policy Coordination 2004). The ideas behind these new structures are not entirely new (Gray and Sanders 2006: 19–21). Under the previous Labor Government for example, Daffen proposed that the Department of Administrative Services should take on just such a coordination role at national and community levels (Daffen 1995: 21–5, 39–51).

The complementary question is that of what scope Aboriginal organisations will be allowed to determine use of those moneys. The mode of transferring funds counts for much. I will distinguish here between block grants and pooled funding. By the former I mean a grant from a single funding source concerned with a particular functional area, such as health or education, in which the budget consists of one instead of many line items. The Aboriginal organisation is simply presented with, say, $5 million for housing, rather than $5 million divided into smaller amounts for particular structures, landscaping, fittings etc. The funds take the form of a single undifferentiated grant for unified Indigenous management and disbursement on housing needs. In that way, external transfers can, beneath that carapace, be converted into internal operational autonomy.

In its expansion across the country from the 1970s to the 1990s (Sanders 2005: 205–7), the Community Development Employment Projects (CDEP) scheme pioneered a new pattern of block grants in which the funds consolidated are not existing organisational grants, but individual welfare entitlements. By pooling unemployment benefits supplemented with amounts for capital and administration, these schemes place substantial resources at the disposal of Aboriginal management boards that function, generally, at about the suburban or municipal levels, and are able to mobilise the labour of between dozens and hundreds of local participants. At the time of writing, CDEP has passed through a phase of retraction under the policies of the Howard Government. Profitable settlement enterprises, especially general stores and alcohol canteens, offer, more indirectly and mostly at a smaller scale, the same potential. They can convert the transfers of government welfare payments to individual local Aborigines into a substantial retailing profit, some of which can similarly be made available for collective purposes such as airstrips or grants for education or ceremony (Arthur 1999: 7; Commonwealth of Australia 2000a: 90).

By pooled funding I mean the combining of allocations for a particular functional area from a number of funding sources, say a Commonwealth and its corresponding State/Territory department, into a joint payment, such as those disbursed by the Indigenous Housing Authority of the Northern Territory (Commonwealth of Australia 2000a: 65–6). Once again, the aim is to enhance self-determination by allowing a greater degree of internal operational autonomy

(Smith 2002: 20–2). Such arrangements were trialed in the area of Indigenous health.

> Under the trials each of the [Commonwealth, State and Territory] jurisdictions involved contributes an amount to the funds pool, based on an estimate of what would have been available to the community had there not been a trial. . . . .

> This pool of funds, which is no longer constrained by the specific rules of the programs of origin, can then be used by the community for service substitution and to address the health priorities that the community itself considers to be the most pressing.

> The advantage of the trials is that there is much more intensive community involvement, both through consultation prior to the commencement of the trial and in the decision making on health services delivery during the course of the trial (Commonwealth of Australia 2000a: 30).

Conditions 1 and 2: Articulation, aggregation and autonomy

This mechanism has appeared particularly apt in company with the kind of vertical political development discussed previously. Prior to the abolition of the Federal-level ATSIC, a good deal of reformist thinking was focused on redesigning the political structures of Aboriginal affairs towards institutionalising Aboriginal authority at higher levels, and consolidating the revenues passing through those carapaces into Aboriginal domains at once larger and more functionally versatile. Reeves (1998), in his model for a new Aboriginal land rights regime, proposed a Northern Territory Aboriginal Council that would receive all royalties[3] payable from resource developments on Aboriginal lands, to be disbursed through a competitive granting system to regional land councils, and then used to meet local needs for social or economic development.

Centralising the royalty stream would have required a proportion of it to be re-appropriated from existing local royalty associations. These bodies, and their individual members who in some cases receive cash distributions, would have no further direct entitlement (Reeves 1998: 361–3, 609). Reeves hoped that the vastly larger funds directed by ATSIC and Territory agencies into Aboriginal programs would be delivered the same way. Similarly, the Commonwealth Grants Commission raised the possibility of Indigenous-controlled State-level bodies through which all Commonwealth, and perhaps State, funding for Indigenous purposes would pass (Commonwealth of Australia 2000b: 59–60). Such a unified point of articulation in each State or Territory would constitute a single carapace

---

[3] Strictly speaking, the moneys received by Aboriginal interests are royalty equivalents—that is an amount paid to them by government equal to the amount of royalties received by government.

over what, it was argued, would be a more functionally integrated Indigenous domain.

Evident in some of these proposals, however, is a move beyond block and pooled funding to something in the nature of general revenue, in which there is a consolidation of funds from all sources, whether different agencies or levels of government, into a single sum, and no prescription even as to the 'portfolio' area to which the moneys are to be applied. As the terms used here imply, the degree of internal autonomy that this facilitates approaches the prerogatives of government. Pearson (1999, 2000a: 70–2, 78–80), operating at the regional level of Cape York, has argued explicitly that the political relationship between the Aboriginal and non-Aboriginal domains should be negotiated across an interface, in my terms a point of articulation, instituted at regional level, and across which resources should pass from the state with minimal regulation. As part of his campaign to convert 'negative' into 'positive' welfare, these resources should include those moneys currently being provided directly to individual Aborigines as pensions and benefits, thereby replicating the CDEP mechanism at a geographically and financially more inclusive level (Pearson 2000a: 88).

A further possibility arises from the work of Pretes (2005), who reviewed national and provincial-level trust arrangements in other parts of the world. These convert windfall or limited-term resource development payments into a permanent income-producing trust fund by means of investment in metropolitan capital markets. He reports that the tiny Pacific island country of Tuvalu entered an arrangement with aid donors for the advance payment of aid moneys into a trust, the income of which has largely removed the need for aid (Pretes 2005: 230–42). East Timor's Petroleum Fund is another example. Such arrangements have not been widely explored at the local or regional Aboriginal level, but they signal the possibility of a further step, from the device of converting external subsidy into internal operational autonomy, to transforming it into dedicated capital (eg. Altman 1985a: 143–5 and passim).

By progressively reducing the degree of prescription that accompanies external funding as it enters the Aboriginal domain, such institutional devices turn the organisational structure into a more absolute transactional boundary, and a more effective carapace. However, it is only necessary to recount these proposals of several years ago to be struck by how sharply they run against the more recent trend of policy. While that trend is now indeed towards program coordination and pooling of funds, developmental goals are to be negotiated between Indigenous interests and government, and progress towards those goals subject to official monitoring (Office of Indigenous Policy Coordination 2004). Innovations such as the health funding trials were among the last to explicitly endorse what was in effect a self-determination ethos at the level of local application.

Following the abolition of ATSIC, renewed effort was directed at the coordination of administration and funding both between agencies and across levels of government in all areas of Indigenous servicing, as mentioned above. This 'whole of government' approach was combined, however, with a new assertiveness by Federal policy-makers insisting on the achievement of negotiated outcomes aimed at the improvement of conventional social indicators. This new attitude was proclaimed primarily by the highly publicised instrument of Shared Responsibility Agreements, although they are supposed to cover discretionary funding only (Commonwealth of Australia 2005: 100–3; McCausland 2005a), and so far account for a tiny proportion of spending on Indigenous affairs (Gray and Sanders 2006: 13–14).[4] Most recently, the Federal intervention in Northern Territory Aboriginal communities has dramatised government insistence on integration and public oversight as the path to acceptable outcomes in the policy arena of remote area Indigenous development. These initiatives were undertaken without reference to the major Northern Territory land councils. In other words, Aboriginal organisations are to become less of a transactional boundary, not more. The carapace is being removed.

Condition 3: Alternative sources of supply

Having considered up to this point structural conditions affecting the Aboriginal organisation in its relations with the external world of non-Aboriginal resource providers, I turn now to look at its relations with the Aboriginal domain that it services and represents. In this section I briefly discuss a third condition affecting the capacity of organisations to bring about self-determined development. This relates again to the extent to which an organisation is able to satisfy Rowley's image of the carapace, a protective layer that intercedes between the Aboriginal domain and the outside world, manages incoming traffic from non-Aboriginal agencies, and creates interior space for the formulation of Aboriginal priorities and responses. The limiting factor here is the availability of alternative service providers, and the scope that individuals and families from within the Aboriginal domain have, or seek, to conduct their affairs through alternative connections. The capacity of Aboriginal agencies to effect developmental change is dependent upon the relative intensity of their engagement with local Aboriginal affairs, with the daily functionality of Aboriginal lives, both in itself and by comparison with the Aboriginal business that passes them by. A comparison may help to make the point.

Remote area organisations seem best placed to maximise their share of dealings with Aborigines within their particular geographical and functional catchments. So, for example, when Pearson has made his arguments for a regional Cape York

---

[4] Gray and Sanders (2006: 13) refer to 120 Shared Responsibility Agreements signed by November 2005. The government website <http://www.indigenous.gov.au/sra/search/srasearch.aspx> lists 267 as at December 2007.

planning process that will take charge of all resources due to Aborigines from government and other parties, he then turns to the question of internal allocation for social and economic purposes beneath an Aboriginal organisational carapace that can claim almost comprehensive coverage of Aboriginal affairs on the settlements within that region. His proposals on this, as evidenced in his distinction between 'positive' and 'negative' welfare, are about changing the impact of existing resource inputs by managing them in a manner that promotes development instead of perpetuating dependency. Pearson's vision takes advantage of the conditions of remote Australia, where the lines of articulation connecting Aboriginal people on the ground with external servicing and welfare regimes, or to sources of supply in general, are limited in number, and therefore able to be negotiated and rationalised into a unified regional management structure. Perhaps the most ambitious instance of such a program was that attempted by the Indigenous Nodom and Pindan companies in the Pilbara in the 1940s and 1950s, in which company governance aimed to provide an exclusive regime of social control and development for all Aboriginal camp residents (Holcombe 2006: 7, 9, 11).

Turning to settled Australia, such possibilities for consolidated management are more limited, and were explored mainly at the local level through CDEP schemes, most now abolished. Whatever Aboriginal carapace can be constructed in the settled regions, it is always partial in both functional and geographical terms, existing alongside a multiplicity of non-Aboriginal agencies—shops, government offices, charities, employers, schools—through which Aboriginal people are able to conduct their affairs. The following passage describes the reach of Aboriginal organisational domains, and intersperses them with those mainstream domains accessed through Aboriginal linkage staff, in the small New South Wales country town of Walgett as listed by a member of a local Aboriginal family.

> In the town there are several Aboriginal community services. . . . Today the most central services are the Walgett Aboriginal Medical Service and the Aboriginal Lands Council. There are also Aboriginal employment services and schemes such as the Community Development Employment Program. Also important are Aboriginal education services and committees on three school campuses in the town . . . and the two Aboriginal Education Officers based at the Walgett branch of the NSW TAFE department. Walgett also has an Aboriginal Legal Service, several Aboriginal Health Workers and an Aboriginal Nurse at the Walgett District Hospital, Euragai Goondi which is a home for the elderly, operating also as a conference centre and accommodation service, Aboriginal Police Liaison Officers, Aboriginal Meals on Wheels workers, an Aboriginal community worker for the Department Of Community Services, three Aboriginal football teams, an Aboriginal Lawn Bowl team, and Aboriginal Cricket, Golf and Darts teams (Peters-Little 2000: 12–13).

With the exception of rural property holdings, there is also no bounded or coherent physical domain. Aboriginal lives in settled Australia proceed in contiguity with an existing civic society. Aboriginal organisations, on the one hand, can offer only patchy coverage for people wishing to satisfy their needs by dealing with other Aborigines, and on the other hand, are less able to enforce social disciplines or participation in developmental programs. There are too many competing lines of articulation, which prevent the Aboriginal carapace in settled Australia from taking on the dimensions of broad civic authority that Pearson proposes for Cape York, and which allow its functional reach to be evaded.

That, indeed, is an explanation advanced by Sanders, Taylor and Ross (2000: 14–15) for the pattern of voter participation in ATSIC elections. Participation was highest in the remote ATSIC regions, lower in the southern settled regions, and lowest in capital city regions, because, they suggested, ATSIC resources became progressively less significant in each type of area as a proportion of total servicing resources and market opportunities available, and thus ATSIC was of correspondingly less importance in Indigenous peoples' lives. The same could be said of Aboriginal organisations in general. Conversely, in the remote Maningrida area of north Arnhem Land, the major improvement in infrastructure and amenities provided to the region's outstations by the local Aboriginal corporation was 'underwritten primarily by ATSIC in the 1990s' (Altman and Johnson 2000: 11). The contrast between remote and settled Australia thus suggests that the potential for an Indigenous model of development to emerge beneath the Aboriginal organisational carapace is inversely related to the extent of alternative individualised articulations maintained outside that carapace.

Condition 4: Organisations and 'community'

To this point, the ability of Aboriginal organisations to foster an Indigenous model of development has been discussed in terms of an interaction between their functions as a point of articulation and as carapace, that is, their capacity to operate as a transactional boundary within the policy and program structures of Aboriginal affairs. Other considerations emerge from looking beneath the carapace at the question of the internal coherence and integrity of the Aboriginal domains that they represent (Mantziaris and Martin 2000: 281–6). It directs our attention to endogenous constraints on developmental possibilities. As higher levels of organisation will inevitably comprehend more diverse constituencies, sometimes mutually hostile ones (Tonkinson 1984-5: 381), this question is sensibly pursued at the local scale.

Rowley appears to have regarded the relationship between community and organisation as unproblematic, perhaps organic. Sullivan, as evidenced by the title of his discussion paper 'A sacred land, a sovereign people, an Aboriginal corporation' (1997), similarly takes as fundamental the need for land-controlling organisations to faithfully embody a social community. The histories of many of the remote settlements for which formal incorporation was first conceived, as artificial clusters of traditionally separate and widespread groups, might have caused some prima facie reservations about the representative capacity of any single administration. Still, settlement councils and the like sprang from the realities of residence, and so could claim at least a secular municipal realism, if not a cultural authority (see Sullivan 1996: 10–12; Smith 2002: 23–4). The same can be said for some legal, health and other services that have been organised from the ground up in response to a localised Indigenous need. Holcombe (2004b) has gone further to describe how co-residence at a remote settlement can indeed provide a basis for the development of real communality among people of diverse traditional attachments.

In other cases, organisations have been created less as the incorporated form of a discrete pre-existing Indigenous domain and more as a contrivance around some connection between Indigenous space and an external contingency that happens to have policy relevance. Northern Territory royalty associations, for example, are constituted by s.35 of the ALRA around the accidental geography of a mineral deposit (Altman 1983: 126–30; Levitus 1991). Similarly accidental are the boundaries of land left available for transfer in various jurisdictions, but which make up an immutable frame within which Aborigines, whose attachments may extend into the land from different directions, must organise themselves as claimants, and then as owners and managers.

Policy initiatives, such as the devolved New South Wales local land councils, can be introduced into places in which colonial history has already embedded a potential for conflict between Indigenous groups. In many parts of Australia, both settled and remote (Macdonald 1997; Martin 1997; Pearson 2000a: 65), the localised distinctions between 'traditional' and 'historical' peoples structure internal Aboriginal conflicts that bear upon organisations and the disposition of their resources. Similarly, a different cultural consciousness and lack of mutual sympathy may divide town-dwelling 'half-castes' from reserve-dwelling 'blackfellas' (Sullivan 1996: 86–90). These common conditions of the political genesis of organisations condition also the ongoing character of political business conducted around and within them. Disputes among native title or land rights claimants, family politicking for capture of the resources of local organisations, disquiet over the introduced artificiality of 'communities' and the legitimacy of those deemed to be politically representative, all go to subvert any preconceptions of communitas beneath the carapace (Brennan 1998: 34–5; Macdonald 2004: 43; Mantziaris and Martin 2000: 277–80; Peters-Little 2000: 13–14, 17–21; Rowse 2000b: 1525). They therefore also bear upon any prospects for the endogenous mapping out of new paths of development.

Condition 4 (cont.): The political economy of organisation and domain

We can, then, presume no necessary and natural dimensions of correspondence between Aboriginal groupings and Aboriginal organisations. However, recognising the potential for conflict and schism that resides in the historical or organisational overriding of Aboriginal diversity does not exhaust the issue. We can take it further by looking at two studies from the early 1980s that were not focused specifically on Aboriginal organisations, but posed alternative conceptions of the remote-area Aboriginal domain as an object of development policy.

Elspeth Young's (1981) portraits of the Northern Territory settlements of Yuendumu, Willowra and Numbulwar sought to forge a link between the recognition of development problems confined within economics, and the recognition of culture confined within anthropology. Young saw Aboriginal communities as a domain of difference. That difference affected both in degree and in kind their need for housing, health, education and employment, and affected also their capacity to derive benefit from the prevailing servicing regimes for all those things. Developing the Aboriginal economy demanded an understanding of that economy as Aborigines experienced it, including their understandings of money, commodities, welfare, land and business enterprise (for example, Young 1981: 162–4). Only on the basis of those levels of understanding could self-determination be realised. Young wanted policy to reform itself so that it could assist 'in bringing about the development of lifestyles closer to those which Aborigines desire' (1981: 3). Illustrative of her appreciation of Aboriginal lifestyle and thus of the internal character of the Aboriginal domain was her view of the implications of a shortage of cash at Yuendumu in 1978, when in one month only 41 per cent of adults received a significant monetary income.

> Thus more than half of Aboriginal adults resident in Yuendumu have to depend on the generosity of their relatives and friends for money to buy food, clothing and other necessities. While the sharing of these resources is an integral part of the social system, vital to the maintenance of status and consolidation of alliances and relationships, the amount available for redistribution, as indicated by the low per capita income, is clearly less than required (Young 1981: 109).

Young thus interprets the management of this money shortage in the positive terms of cultural consolidation. It appears, however, to indicate major shifts in the politics of sharing. Subsistence resources were no longer sufficient for family needs, and access to them was rationed by controllers of vehicles, so the broad distribution of production rights and the balance of exchange linkages associated with that domain of bush provisioning were on the whole less significant for

Yuendumu people. Cash income was now needed, but its distribution was substantially unbalanced. More than half the adult population was thus critically under-resourced for participation in the politics of reciprocity and patronage. Young's remarks suggest an intensive and unilateral deployment of kinship as a currency of supplication, which means that some people with money income were engaged in the constant tactical fielding of demands. We do not know to what extent this was burdensome or empowering (Austin-Broos 2003: 125-26; Peterson 1997: 190).

John von Sturmer studied the impact of major resource developments in western Arnhem Land from 1979 to 1983. He also saw difference, not in the sense of generalised cultural preferences, but as alternative sources and manifestations of power. He recognised how the social values contingent upon connections to country came to depend less on the shifting politics of access to increase sites and their bounty (von Sturmer 1984a), and more on the fixity of rights to mining sites and their royalties. He showed also how they had been joined by new structural points of control over resource flows into Oenpelli and Kakadu National Park. Von Sturmer (1984b: 151–63) thus probed the implications of self-determination policy and mining development in terms of the changing niches and rewards of employment and residence open to the Aborigines of western Arnhem Land.

The irony in this comparison lies in the relationship that each scholar perceives between Aboriginal agency and development. Young represents Aborigines as respondents of policy, and wants policy amended in ways that will allow Aborigines to respond to more desirable effect. Von Sturmer sees Aborigines as active calculators to whom policy presents opportunities for strategic emplacements leading to personal advantage (see also Gerritsen 1982: 21–6). Altman and Smith's (1994) subsequent study of the Nabarlek Traditional Owners Association clarified the anti-developmental outcomes implied by von Sturmer's analysis. Almost all moneys paid to local Aboriginal interests from the Nabarlek mine were disbursed in the form of vehicles and cash distributions, to beneficiaries determined more by personal networking than by traditional entitlement or social need. The amount devoted to social or financial investments that could produce lasting benefits was tiny. In contrast to Young, then, von Sturmer recognised not just a domain of difference, and not a moral economy, but an Aboriginal political economy.

These perspectives are relevant here for how they allow us to perceive Aboriginal organisations and the Indigenous domains that they overlook. Young's perspective acknowledges an Aboriginal desire for development of a culturally appropriate kind, but does not recognise internal differentiation in terms of the capacity of particular individuals or families to control traditional or introduced resources, and to interpolate themselves between those resources and other people who need them. Her perspective equips us to understand Aboriginal organisations that service such domains and claim to speak for them, in similarly opaque terms—that is according to their self-representation as facilitators of self-determined development. To that extent Young validates from the field the expectations of Rowley and Coombs as to the likely character of Aboriginal organisations as a policy intervention.

Von Sturmer's perspective makes us look at that policy intervention differently. By investing Aborigines with a political capacity and relating that capacity to new and old sources of power, he asks us to analyse the impact of the Aboriginal organisation in terms of structural innovation and the internal redistributions that flow from it. In other words, von Sturmer's approach points to a complementarity between the two policy functions of the Aboriginal organisation. As a point of articulation it channels material resources, and as a carapace it offers a set of political niches that are funded by, and positioned to take advantage of, that resource flow.

In addressing the relationship between Aboriginal organisations and domains, we therefore need to think in terms of a realpolitik of articulation. The picture is one of web-lines of major and minor fractures in the Aboriginal polity fanning out from every point at which resources feed in from the non-Aboriginal world. It requires us to place inverted commas around the word community, and offers a framework for understanding the excess of small-scale incorporations experienced in some areas (Commonwealth of Australia 2004: 130–31; Smith 2002: 9). It makes a matter for analysis the question of what new points of articulation and an enhanced flow of resources cause to happen among Aborigines and the implications of these things for development, and it moves such analysis beyond ethnic boundaries and the politics of difference to re-focus on older concerns like kinship, patronage, corporate groups and power (Austin-Broos 2003; Folds 2001: 144–7; Macdonald 2000: 106–9; Mantziaris and Martin 2000: 192–4; Smith 2005).

The implication here is that the policy conception of Aboriginal organisations as protagonists on behalf of an agreed community conception of welfare and progress must take cognisance of the standing of those organisations in relation to Aboriginal domains that are themselves political fields. These fields may include, it should be said, some political actors who genuinely seek to advance just such a common conception of welfare and progress. Thus, we need to remain open as well to the possibility that constructive organisational leadership can actually overcome family factionalism, as Pearson (2000a: 69–70) claims has occurred in the town of Coen, but that too must be a matter for analysis, and not accepted prima facie as a connotation of the term 'community'.

Aboriginal organisations and mining

In a limited number of cases, a large proportion of the resources flowing to Aboriginal organisations come from agreements negotiated with resource developers. In such cases, that financial point of origin has implications for the way in which the four structural factors identified here bear upon those organisations.

The political level at which mining revenues enter the Aboriginal domain will depend on the legislative and institutional framework for managing them. In the Northern Territory, the formula given by the ALRA for distributing mining royalty equivalents from Aboriginal land for long directed the largest share (now somewhat reduced) to the Aboriginal land councils (Altman and Levitus 1999: 7–8). The two largest of these operate as high-level political and administrative organisations. In the 1980s Altman and Dillon (1988: 126) observed that

> land councils represent the most advanced and comprehensive expression of Aboriginal self-government in the NT (and for that matter anywhere in Australia). Land councils' activities are increasingly para-governmental in nature.

That such a level of Indigenous political authority is funded from resource development is, of course, a peculiarity of the land rights legislation created for the Territory in the 1970s. Such an arrangement has not been replicated in other jurisdictions, and in its absence, the mining industry has found no reason of its own to fund Aboriginal interests at that level for anything beyond discrete projects of mutual concern. While mining companies may be international entities, their operating units are always local, and the business case for protecting their social license to operate by fostering Aboriginal development will be focused on Aboriginal groups in the areas affected by mining. Indeed, this local imperative is written into the Northern Territory legislation itself, under which traditional owners have veto rights over proposed exploration. Thus, the uranium company Energy Resources of Australia is legally required to deal with the Northern Land Council, but it has nurtured a relationship with the local Gundjehmi Aboriginal Corporation, the members of which are primary traditional owners of the Ranger and Jabiluka uranium deposits.

Even in the absence of a veto, hostile Aboriginal political action is bad both for its impact on public opinion towards the industry, and as a direct cost in the form of an impediment to land access or project continuity (Harvey 2002b: 4; Hawke and Gallagher 1989; Trebeck 2007a). For such reasons, mining companies will generally seek to concentrate resources flowing to Aboriginal interests at a local level. Consequently, and especially in a policy environment that has in recent years abandoned concern with high-level Aboriginal political development, the point of articulation across which resources flow from mine operators into Aboriginal domains will tend to be occupied by organisations constituted within limits that are proximate to the extraction project itself.

While that local focus is by definition a confining factor, the mode of transfer of resources into that domain can be relatively free. Altman and Dillon (1988: 134) comment that mining moneys in the Northern Territory 'from the Aboriginal perspective . . . are the single most significant source of finance that is untied and that can be used for development purposes'. Local associations established to represent people in the areas affected by mines on Aboriginal land are entitled to at least 30 per cent of the royalty equivalents payable from those mines.

Outside the Territory, Rio Tinto is pioneering a similar transfer of substantial resources to local Aboriginal control. In the East Kimberley, one of the criticisms directed by Aborigines at the notorious Glen Hill Agreement and the subsequent Good Neighbour Agreement entered into by the operators of the Argyle Diamond Mine in the 1980s was that the company controlled all moneys and decided what benefits were to be provided to the recipient communities (Dixon 1990). Moneys distributed by the Argyle Social Impact Group were similarly subject to non-Aboriginal criteria (Howitt 2001: 244–5). When the company moved into a new generation of community relations thinking and signed the Argyle Participation Agreement in 2004 (Harvey and Brereton 2005: 11), funding was directed to a number of community-controlled trusts.

In western Cape York, Comalco established the Weipa Aborigines Society in 1972 to carry out projects for Aboriginal benefit. Its lack of accountability to Aboriginal interests led to the creation of the locally operated Napranum Aboriginal Corporation in 1993, but corporate funding conditions continued to restrict the independence of that organisation as well (Howitt 2001: 220–2). Under the Western Cape Communities Co-existence Agreement negotiated between 1996 and 2001, a minimum of $4 million annually is transferred from the company and the Queensland Government to the Western Cape Community Trust, under the majority control of local traditional owners (Harvey 2004: 241–42).

While such funding transfers under mining agreements are now less likely to allow for unrestrained and immediate consumption expenditure (Altman and Smith 1994; ISS/ACIL 2001: 37), to the extent that they are not accompanied by any specific spending prescription they preserve a potential for self-determined development. Such arrangements respect the status of the recipient organisations as a transactional boundary, and thus contrast with the ideological shift in the general Indigenous affairs policy environment represented by Shared Responsibility Agreements. Indeed, Rio Tinto's senior Community and Aboriginal Relations officer was admonished by a public service bureaucrat that the relatively untied money provided by miners made the task of promoting Shared Responsibility Agreements more difficult for government (B. Harvey pers.

comm.). Where government funding is increasingly tied to agreed outcomes, mining agreements may thus offer a preferred and relatively independent operating environment for Aboriginal organisations.

With respect to the third structural condition discussed above, that is the availability of alternative providers with which an Aboriginal organisation's constituency may prefer to conduct their affairs, the significance of mining revenues seems ambivalent. Those revenues, often being of the scale of a windfall for local Aborigines and relatively untied, can appear to present the promise of a coherent endogenous strategy of social development being designed and applied within a local Aboriginal domain beneath the protective coverage of a complete organisational carapace, much after the vision of the Nodom and Pindan companies mentioned earlier. Indeed the practice of substitution (whereby government agencies direct discretionary funding to communities without access to mining-related benefits and thereby force those groups funded by mining to use their income as a substitute for government servicing) can push onto such organisations a wider responsibility for local servicing than they may wish to have. But at the same time, major resource development projects in remote areas improve transport and communications and occasion the implanting of enclaves of western civil society alongside. In western Cape York, north-east and western Arnhem Land, the Pilbara and elsewhere, towns exist because of mining. The servicing agencies that populate these places can offer such a multiplicity of contacts to local Aborigines that their combined impact is not only confusing but anti-developmental, and their rationalisation has lately become a major priority in Indigenous policy (Edmunds 1989: 79–81; Gray and Sanders 2006).

Finally, the relationship between organisation and domain can be a particularly vexed feature of the structural context. I referred earlier to Northern Territory royalty associations as an example of organisations built upon an external contingency, their memberships determined by some process and politics of recognition of Aboriginal spatial relationships with the mine site. Any sizeable Aboriginal corporate entity will include categorical distinctions within its membership that can provide the raw material for political fission. Organisations that exist to receive benefits from a mine, however, would appear, by the absence of any 'natural' constituency and therefore of any original legitimacy, to be unusually predisposed to that risk.

I have elsewhere explored the impact of this failure of collective commitment in the history of the Gagudju Association of Kakadu National Park (Levitus 2005). Its replacement as the royalty-receiving association for the Ranger Uranium Mine by the Gundjehmi Aboriginal Corporation elevated narrowly conceived traditional rights over the mine site at the expense of a more expansive, though incoherently realised (Levitus 1991), concept of affected peoples. Similarly, the Rio Tinto officer mentioned above has remarked that one of the dangers for the company's policy of promoting regional Indigenous development in its areas of operation is the tendency in some circumstances for every clan and family to demand that the company pays for a lawyer to negotiate their own separate deal (B. Harvey pers. comm.).

In summary, then, the distinctive potential for sustainable and self-determined Aboriginal development offered by remote-area mining projects arises from their capacity to bring a large and relatively untied revenue flow to bear within a limited geographical range, through the agency of a locally constituted organisational structure. That potential, then, flows from the standing of such organisations with respect to the first two structural conditions discussed in this chapter. Perhaps that will not be enough. The record of the Gagudju Association suggests as much, for surely a local operational focus and active constituency and a large untied revenue flow were two of its strengths. Von Sturmer is one observer who requires more. The structural conditions that he listed for the success of west Arnhem royalty-receiving associations were akin to the third and fourth discussed here:

> Success will hinge on their capacity to incorporate and to reflect all the features of Aboriginal social structure and organisation; to deflect fissive pressures; and to incorporate all other organisations operating within their sphere of influence within a single hierarchically-ordered structure; . . . If they are to work, the associations must constitute literally a system of local government and be able to ensure social equity. They must be able to make rules, and to enforce them (von Sturmer 1982: 99–100—his emphases).

Clearly only some aspects of these specifications are controllable though negotiation and agreement with the miners. They evoke again the image of the complete carapace, and imply at least a political unity, if not Rowley's organic identity, between organisation and domain. Organisations established on this vision, even remote ones, will always be heir to the challenges of alternative external connections and internal differentiation.

The assimilation of resources

After acknowledging the structural limitations discussed so far, we need also to recognise the strategic centrality of Aboriginal organisations. This discussion has used two characterisations: each organisation occupies a point of articulation, that is, a bridge for the movement of resources across that intersection of policy and culture, and each is a carapace, marking out a functional framework beneath which a more-or-less coherent Aboriginal domain of action and values can be created, enhanced or preserved. As Rowley recognised in the beginning, that duality offers the potential, even within a context of dependency, to exercise some Aboriginal influence over processes of change.

One important function that local Aboriginal organisations can have is to deflect and manage the impact of external subsidies upon Indigenous recipients. It is a function that has been most obviously and commonly exercised by CDEP schemes, that intervenes in the previously direct flow of money from social security providers into consumption expenditure by the clients, consolidate those multiple individuated flows into a resource and channel it through projects for community benefit. Northern Territory royalty associations illustrate other features of this intervening role. Royalty associations receive shares of the payments made from resource development projects under the ALRA. During the extended debate over reform of this legislation following the Reeves Report, the question arose as to whether there was any need to prescribe Aboriginal structures of decision-making below the level of a regional land council, that is, whether local royalty associations had any role to play. The view held by Reeves (1998: 207, 596) and others, that local traditional processes could be allowed free play, was in my view misconceived (Levitus 1999: 126–7). This issue goes to understanding the character of local organisations as, in the terms used above, a formal transactional boundary and a conduit for the transfer of resources, between the non-Aboriginal world and the organisations' memberships.

There are two important respects in which the allocatory tasks of local associations are different from those managed in pre-colonial times within traditional Aboriginal society. Firstly, the bundles of rights and benefits at stake are of enormous quantity by comparison with anything that passed through hunter-gatherer processes of production and distribution. The material values represented in a government instrumentality's one-line budget item, or in a single payment of mining royalty equivalents, is vast and unprecedented by comparison with the volume of desirables handled by traditional allocatory systems. Secondly, and perhaps more importantly, these benefits and entitlements do not physically originate from within the Indigenous domain, and so are not already enmeshed in those relationships of reciprocity and obligation that attach to locally produced items. Rather, they have an external source and are arbitrary both as to their timing and point of entry into the Indigenous system. They are unsocialised goods. So a program is approved and a delivery arrives—a housing project, a convoy of Toyotas, an office equipped with modern communication equipment—and local processes have to absorb it.

Because these items are so substantial, and management of them within the Indigenous domain is not already conditioned by relationships embedded within the circumstances of their production, it is necessary to design mediating structures and processes that function not to replace, but to assimilate them into, locally self-determined distributory mechanisms. That should be the function of royalty associations and like organisations. This connects with some of the concerns discussed above and with many of the considerations raised in discussions of governance and internal accountability (Dodson and Smith 2003;

Hunt and Smith 2006b). It is complementary to the point made above, about the conversion of external subsidy into internal operational autonomy. That earlier point was about freeing resources from bureaucratic prescription and making them available to Aboriginal priorities. The present point is about subjecting those resources to a structured and considered passage into local Indigenous domains. Both matters need to be understood as preconditions to the implementation of Aboriginal development plans.

Conclusion

Aboriginal organisations were seen as an instrument through which a self-determined Aboriginal developmental trajectory could be formulated and pursued. They were an institutional home for what remained of policy's aspirations for Aboriginal development after that concept was surrendered into the discourse of self-determination in the 1970s. Their capacity to satisfy such aspirations can, from the outside, be seen to have depended on a number of structural conditions, operating both above, with respect to the supervening non-Aboriginal domain, and beneath, with respect to their own capacity as carapace. Those structural conditions, bearing upon the status of the organisation as a transactional boundary, continue to be a focus of policy innovation and discord. Most recently, the trend of policy has been to sidestep or overrule that boundary and project its managerial agency directly into local Aboriginal domains.

I have suggested here that Aboriginal organisations funded principally from mining projects are now relatively well placed to preserve internal operational autonomy. It is clear, despite the changing relationship between the mining industry and Indigenous people in Australia, that establishing and resourcing such organisations is not a widely acknowledged channel for transferring benefits to Aboriginal peoples affected by mining. The Australian Government's 'Working in Partnership' program for developing cooperative relationships between the mining industry and Indigenous communities heavily emphasises training, employment and business development. A ready possible explanation for this flows directly from my previous point. Government is more interested in promoting the mainstreaming linkages of employment and ancillary business between projects and local communities than in resourcing self-determination. Alternatively it may simply reflect a reasonable up-front emphasis on the immediate opportunities for project participation, relegating post-project sustainability to downstream planning.

There also seems to be limited industry recognition. Few of the companies subscribing to the Working in Partnership program provide revenue streams to local organisations under majority Indigenous direction. Indeed, of those mine sites surveyed by the Australian Bureau of Agricultural and Resource Economics that were located in the vicinity of a discrete Indigenous community, half had no agreements of any kind with local Indigenous people at the mining stage of their operations (Tedesco, Fainstein and Hogan 2003: 24). In another survey published in 2001, a highly disproportionate number of agreements between mining companies and Indigenous communities were attributable to only two companies, Rio Tinto and Normandy (ISS/ACIL 2001: 15). However, 'payments or other compensation mechanisms' was the most common kind of provision included in the 140 agreements surveyed, covering 'lump sum payments, soft loans, royalty equivalents, and equity in companies, rents and lease payments' (ISS/ACIL 2001: 17). Not all of these will produce a revenue stream for a locally-controlled Aboriginal organisation, but this figure provides an indication of a potential to be explored. These survey data suggest that few players are including the resourcing of Aboriginal organisations and post-mining development within their range of expectations from the advent of mining, but those players are important ones.

In the long-run history of any remote Aboriginal town facing deteriorating demographic trends (Taylor Chapter 3), the duration of even a major mining project is only a window (Render 2005: 35). Moreover, there will always be many people of working age in such areas who will never achieve the required state of work-readiness, or who have no wish to work in a mine, or who are numerically beyond the capacity of the mining industry to absorb them. In the relationship between mining companies and local Aboriginal populations, the resources and capacities of Aboriginal organisations will be an important determinant of the long-term developmental legacy a mining project will leave behind. The initiatives of such organisations will be a critical mediator between the windfall revenues from mining, and what those revenues are able to create and sustain for the Aborigines in that area. While the remote-area mining industry in Australia is currently undergoing unprecedented expansion, and there will always be major new projects eventuating here and there, for any given Aboriginal community the particular project occurring in their vicinity may be the only opportunity in much more than a generation to acquire the capital base for at least a partially self-resourced future. As O'Faircheallaigh remarks, 'a community will tend to have just one chance to extract revenue; if it fails in relation to this specific project, it fails in total' (1996: 198).

We leave the discussion, then, at the point where it begins to broach another issue, that is, the kind of development that Indigenous peoples may wish to foster. In the international discourse around the relationship between miners and indigenes, some have begun distinguishing between economic development and what is variously called culturally appropriate, community-led, or human, development, thinking less about 'how to fit Indigenous peoples into a commercial mining operation, and looking instead at how a commercial mining operation can fit into Indigenous life' (Render 2005: 35–8). This not only raises new questions of definition and evaluation, but places responsibility on

Aborigines themselves to determine what development should be, and how to bring it about. In this context, the matters discussed in this chapter are structural preconditions, important because they affect the ability of Indigenous interests to equip themselves with the institutional instruments necessary for development with or without mining. They are about using the resource transfer instigated by mining to create a space in which they may consider their own options for development. The work of development itself has still to follow.

5. The governance of agreements between Aboriginal people and resource developers: Principles for sustainability

David F. Martin

Introduction

In February 2008, the Commonwealth Attorney General called for a new approach to resolving native title claims, unblocking the system through 'interests-based' negotiations between claimants and other parties, including governments, which can result in an array of 'non-native title' outcomes (McClelland 2008). More recently, in the 2008 Mabo Lecture the Commonwealth Minister for Families, Housing, Community Services and Indigenous Affairs (Macklin 2008) called for 'a mindset which structures the governance of these arrangements to ensure financial benefits create employment and educational opportunities for individuals and are invested for the long term benefit of communities'.[1]

These Ministerial speeches raise a number of important questions. The first is that they do not engage with research in relation to major Australian mining agreements which argues that many do not in fact deliver substantive, meaningful benefits to the Aboriginal parties (for example, Cousins and Nieuwenhuysen 1984; O'Faircheallaigh 1988, 2006). Even when agreements do deliver such benefits, they would appear to have only minimal impact on the general socioeconomic status of the Aboriginal people in mine hinterlands (for example, Taylor and Scambary 2005).[2] Secondly, the speeches avoid reference to the important issue which arises where compensatory payments involve the substitution of monetary benefits with goods and services which arguably should be generally accessible to all citizens (Smith 2001: 44; see also, for example, Altman 1985b; Altman and Levitus 1999: 17–18; Toohey 1984).

Thirdly, there are questions relating to the 'distributive equity' of agreement payments (that is, ensuring that compensation is paid to the appropriate rights holders) and 'distributive spread' (for example, whether other Aboriginal people than just those whose native title rights are impacted, should also receive a portion of compensatory payments) (Smith 2001: 42). Fourthly, there is the issue

---

[1] Such sentiments are far from novel—they have a history at least as long as that of land rights itself; see for example Turnbull 1978.
[2] Indeed, such factors as the lack of adequate baseline data in many regions, compounded by difficulties in obtaining quantitative data sets that relate to the specific groups who are the beneficiaries of an agreement (Taylor 2008a), make the monitoring of agreement impacts potentially quite difficult.

raised by the commonly inadequate attention typically paid to the implementation and monitoring stages of mining agreements (O'Faircheallaigh 2002a, 2003b). Lastly, and the subject of this chapter, are questions of the governance of mining agreements, which go to the ongoing ability of agreements to sustainably and effectively deliver the negotiated outcomes of any particular agreement. The significance of effective and sustainable agreement governance applies *irrespective* of these preceding factors, but little attention appears to have been paid to it as an issue, although attention has been paid to the governance of royalty associations and the like (for example, Altman 1985b; Altman and Smith 1999).

This chapter examines this core requirement of agreements to meet their objectives, making three key arguments: that inadequate attention is paid to the governance of agreements as *systems*; that agreement governance has to be explicitly understood and implemented as *transformative*; and that agreement governance should be seen as *intercultural*, a characteristic operating differentially across the various entities and relationships within any particular agreement.

These are not the only significant governance characteristics. For example, sustainable governance needs to incorporate an active recognition of the diversity amongst Aboriginal stakeholders, and between them and other parties (see Holcombe, Chapter 7; Scambary, Chapter 8). Indeed, the need to incorporate recognition of diversity is not unique to the case of Aboriginal people; there are arguments that it is an essential component of social sustainability more generally (Martin, Hondros and Scambary 2004; Western Australian Council of Social Services 2000). However, the three identified governance issues constitute case studies illustrating important principles for the negotiation, design, and implementation of large-scale mining agreements between Aboriginal people and major resource developers.

The governance of agreements as systems

Mining agreements are negotiated in complex and contested intercultural fields where the parties (Aboriginal groups, resource developers, and governments) bring potentially quite divergent interests and goals to bear on the negotiations, and indeed to subsequent implementation of the resulting agreement. There are substantial structural power and resource imbalances which disadvantage the Aboriginal parties in negotiations. This is despite the leverage which can be offered variously by the *Native Title Act 1993* (Cwlth) (NTA), land rights and heritage legislation, by corporate responsiveness to Aboriginal pressure (Trebeck, Chapter 6), and by the involvement of Aboriginal representative and advocacy organisations with statutory responsibilities (O'Faircheallaigh 2006; but compare Senior 1998: 9). All parties including Aboriginal people seek advice from a range of such specialists as lawyers, resource economists, environmental scientists, investment and tax experts, anthropologists, and others. While negotiations themselves may include varying levels of involvement by the Aboriginal people concerned (Blowes and Trigger 1998; O'Faircheallaigh 2000; Senior 1998; Trigger 1997b), they and other parties are heavily dependent during negotiations on technical advisors, who overwhelmingly determine the form and structures of the resultant agreements.

This is not in any way to argue against the involvement of competent lawyers and other specialists; they are clearly essential for all parties, given such factors as the sometimes byzantine politics and strategic bargaining involved. There are also particular difficulties entailed in developing a common position amongst sometimes factionalised Aboriginal parties, and the complex legal frameworks (such as contract law, and that of the NTA) against which agreements are set, and the need for precision and clarity in their wording also pose challenges (for example, see accounts in O'Faircheallaigh 2000; Senior 1998). However, a failure to also include a specific consideration of governance issues can, I argue here, militate against long-term agreement sustainability. Similarly to the implementation of sustainable development principles more generally in the minerals industry (Brereton 2003), the development of effective and appropriate agreement governance requires a multidisciplinary approach.

The technical experts for each of the parties usually play a significant role in developing the broad terms and principles of a deal, and subsequently have major responsibility for the translation of that deal into a legally binding document which ensures the rights and obligations of the parties are specified sufficiently that they can be protected or enforced at law (O'Faircheallaigh 2000: 12). Key interlinked issues identified by Aboriginal parties are formulated against a background of concerns about the adverse potential impacts of the proposed development on country and culture, a wish to address the historic injustices which are reflected in continuing exclusion from meaningful participation in regional political and economic systems a desire for recognition of political autonomy and self-determination, a concern for capacity to control and access traditional country, and an often ambivalent recognition that mining agreements offer possibilities for leveraging change in their marginalised circumstances.

These concerns arise from within social fields which while intercultural (see discussion in the next section) are characterised by distinctive Aboriginal worldviews, beliefs and practices in which connections to country and kin are very significant (see, for example, Blowes and Trigger 1998; O'Faircheallaigh 2006; Scambary 2007; Sutton 2003; Trigger 1997b, 1998, 2005). There have been enormous impacts on the original economies of the Aboriginal groups now within Australian mine hinterlands, resulting from a whole range of historical factors including the development of the pastoral industry; the sedentarisation of people in towns and settlements; large-scale mining; and, in recent decades, the introduction of the welfare-based cash economy (see, for example, Edmunds 1989; Taylor and Scambary 2005). Aboriginal economies in these regions still operate through dominant principles such as flexible, opportunistic and immediate-return foraging, strong pressure to share resources amongst kin, and (related to this) an anti-accumulation, immediate distribution ethos (Martin 2008a; Peterson 1993, 2005). Scambary (2007) details these matters in his doctoral thesis, summed up elegantly in its title 'My Country, Mine Country'.

The translation previously referred to is not simply a technical exercise in distilling issues raised by Aboriginal people into the legalese of an agreement (contra O'Faircheallaigh 2006). It entails the rendition of what may be quite different and indeed contested notions of cultural and economic futures (Scambary 2007; Trigger 1997a, 1998) into terms which can be recognised and enforced under Australian law—in a manner entirely analogous to the translation of the nature of Aboriginal people's connections to country under their laws and customs to rights and interests in the 'recognition' or 'translation' space of native title (Mantziaris and Martin 2000). What can with the best of intentions still be a rather impoverished translation entails the possibility of a degree of incommensurability between systems of values. Furthermore, the interests-based negotiation process together with standard legal methods combine to break down complex social realities and processes into defined components, and then to establish putative relationships between them. In this way the often generalised and interconnected matters raised by Aboriginal people in negotiations (such as those outlined previously), are distilled and disaggregated into distinct specific elements for which responsibilities may be assigned to particular parties and resources committed for implementation.

For example, concerns about the impacts of mining on country are translated into mechanisms for the protection of its cultural heritage and environmental values; historic exclusion and ongoing socioeconomic marginality are translated into financial benefits, employment and training, and business development provisions. Equally, desires for return to and control of traditional country are met with provisions such as return of mined areas, divesting pastoral stations in the mining hinterland to Aboriginal people, and support for outstation development. Demands for political autonomy and self-determination transmute to self-management and result in the establishment of Aboriginal-controlled entities such as trustee corporations and representative/advocacy bodies, and Aboriginal representation on the raft of committees usually set up under agreements.

This analysis is not meant to imply that there is an alternative set of more culturally appropriate principles by which mining agreements should be negotiated. By their very nature, agreements involving the Aboriginal people of Australian mine hinterlands are intercultural phenomena which must necessarily involve processes of translation (albeit with some inevitable level of incommensurability with Aboriginal values and aspirations) in order to have the intended *practical* effects within the dominant society's legal, political and economic systems. However, I argue that in the negotiating phase inadequate attention is paid to both the governance of agreements as *systems* comprised of the disaggregated components resulting from the negotiation process, and to the implications of agreements as intercultural institutions. In my view these problems arise in part because the technical experts and the parties they represent are not necessarily alert to the complexities and implications of intercultural issues, including those pertaining to agreement governance.

Through the processes discussed above, both the structure of an agreement—and the institutions which may be set up under it—reflect issues identified and agreed to in principle within a translation space established between the values and aspirations of the Aboriginal parties and those of the other parties through the negotiation process and the legal drafting of the agreement. This can be seen particularly clearly in the structures set up under the Century Mine Agreement.

This agreement was signed in May 1997 between representatives of three native title groups, Century Zinc Limited (CZL), then a subsidiary of CRA Limited, and the State of Queensland. The agreement concerned a proposal to open cut mining of a very large zinc deposit north-west of Mt Isa. The formal signing followed protracted and frequently bitter regional negotiations (Blowes and Trigger 1998; Trebeck 2007a; Trigger 1997b) which were initially outside the ambit of the NTA and involved the broader gulf Aboriginal community as well as those asserting native title in the mine site area, the pipeline route, and the port at Karumba.[3] With a breakdown in negotiations leading to the issuance of s.29 Notices under the NTA, there was a consequent change in focus to negotiating with the registered native title claimants under the Right to Negotiate provisions of the Act. Ultimately, final agreement was precipitated by arbitration procedures commenced in February 1997 with strict timeframes under the Act. CZL was purchased from CRA by Pasminco prior to mine site construction beginning, and the Century agreement was renamed the Gulf Communities Agreement (GCA). In 2002, a cash-strapped Pasminco was reconstituted as Zinifex. The terms of the GCA binds all future owners and operators of the mine, including subcontractors.

Benefits from the agreement are directed to members of the three native title groups wherever they may live, and to Aboriginal residents of designated communities in the region, including Doomadgee, Burketown, Mornington Island and Normanton. The GCA and its associated schedules are structured around specific categories of issues and benefits negotiated with the native title

---

[3] This account is drawn from Martin (1998a) which is based on my own knowledge and experience gained during the period when I was engaged by the three native title groups to assist in implementing the terms of the agreement, from information gathered during a short period of fieldwork as part of the CAEPR Mining Project, and from the information provided in Scambary (2007).

groups, and the structures established to address them. CZL is required to provide employment opportunities; provide training programs to assist Aboriginal people in developing the relevant skills; resource and provide financial resources to assist the native title groups and the designated communities to develop small businesses, joint ventures and contracting opportunities; provide monetary payments for the benefit of the native title group; provide assurances about environmental protection in the project area and about the identification, protection and management of culturally significant sites; and access to and interests in pastoral leases held by CZL for the relevant native title groups (Scambary 2007: 109–12).

The bulk of the community benefits package is delivered to the Aboriginal parties through two special purpose organisations established pursuant to the GCA: the Gulf Aboriginal Development Company Ltd (GADC), and the Aboriginal Development Benefits Trust. GADC is a company limited by guarantee whose members are drawn from each of the three native title groups. Its establishment resulted from a political compromise during the latter stages of negotiations, necessitated by entrenched opposition to the regional native title representative body both from the Queensland Government and CZL as well as a degree of distrust of the native title representative body amongst a substantial proportion of the Aboriginal people of the region. GADC was clearly envisaged as playing a significant role in administering the GCA on behalf of the native title groups, and ensuring benefits are delivered in accordance with the intent of the Agreement. After its incorporation, GADC became a formal party to the GCA on behalf of the native title groups. Its functions include distributing monetary payments from CZL directly to Aboriginal corporations representing the native title groups under a formula specified in the GCA, employing a Liaison Officer to assist the Environment Committee and the native title groups in monitoring the project's environmental management regime, an important coordinating role in employment, training and business development, seeking additional resources from government to assist implementation of the Century Employment and Training Plan, providing advice to CZL on contracting and joint venturing possibilities, facilitating the appointment of native title group representatives on the various committees organisations, and assisting in the monitoring and review of the Agreement (Martin 1998a; Scambary 2007: 109–12).

The primary function of the Aboriginal Development Benefits Trust is to promote economic development for members of the native title groups and residents of the designated communities through providing loans or grants for business skills training, start-up finance for small businesses, and financial equity in joint ventures and land purchases. The Trust's principal funds derive from CZL under a formula set out in the Agreement, and at the time the GCA was signed were expected to total around $20 million over the then anticipated 15-year life of the mine. The Trust is also responsible for delivering a relatively small program for sporting development with funds provided by CZL and Queensland.

Other special purpose organisations and committees were established under the GCA. Various pastoral property holding and management companies were set up to enable the gradual transfer of ownership and control of five hitherto CZL-owned properties to the relevant Aboriginal groups—three to Waanyi people, and two to Ganggalida. The Century Employment and Training Committee is an advisory body with representation from each of the native title groups, the designated Aboriginal communities of the gulf, CZL, and government. It has a key role in developing the Century Employment and Training Plan which, along with the Aboriginal Development Benefits Trust, lies at the heart of the GCA's economic development mechanisms. As well, an Environment Committee was set up with the intended role of providing direction to the environmental Liaison Officer and acting as a clearinghouse between the Aboriginal parties and the mine on environmental matters. Finally, a Liaison and Advisory Committee was established for general liaison, reviewing project plans and operations, and as a conduit for information exchange between the project (see above comment) and the members of the native title groups, and designated communities.

Fig. 5.1 Gulf Communities Agreement structure

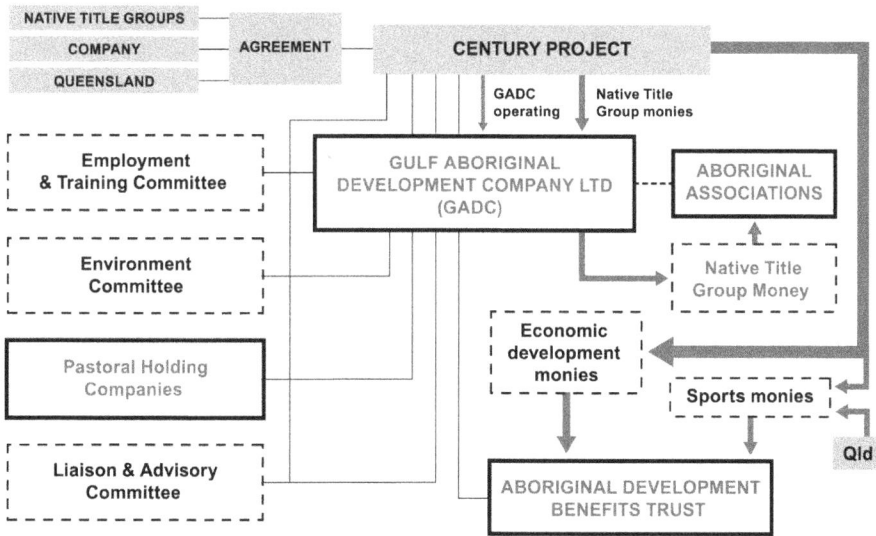

Key: ➡ Financial flows (arrow widths are proportionate to the amount of money involved).
—— Contractual relationships/linkages
Grey text indicates entities that are under Aboriginal control or majority.
Solid box borders indicate incorporated entities, while dashed box borders indicate funds and unicorporated entities.

Source: Martin (1998b), Scambary (2007).

The various entities established under the GCA and aspects of the connections between them are shown in Fig. 5.1. It illustrates an archetypical case of the isomorphism discussed previously, in which responsibilities for particular issues and processes are mapped onto function-specific entities. The viability and effectiveness of this complex and interlinked set of entities is essential to the successful implementation of the Agreement, including delivery of benefits to the Aboriginal parties, as well as the certainty and risk management sought by the resource developer (Pasminco, The State of Queensland and GADC 2002: 7). O'Faircheallaigh (2002a, 2004b) has strongly argued that the implementation phase is given completely inadequate attention in Australian (and Canadian) mining agreements. The GCA provides an instance of another crucial implementation issue to which inadequate attention is paid in negotiating and designing Australian mining agreements—the governance of agreements *as systems*. This is a first order implementation issue because no matter how advantageous to the Aboriginal and other parties a negotiated set of outcomes may be, unless there is sustainable, coordinated, robust and effective governance across an agreement as a whole, there will be a strong likelihood of it failing to meet either mandated outcomes or the expectations of the parties, particularly the Aboriginal ones. This is in fact what occurred in the case of the GCA.

The first five-yearly review of the GCA (Pasminco, The State of Queensland and GADC 2002) found that its institutional arrangements have proved to be uncoordinated, unwieldy, inefficient and far more resource intensive for all parties (in both human and capital terms) than was originally envisaged when the agreement was negotiated. The requirement for Aboriginal representation on the plethora of boards and committees has placed particular strain on the subset of Aboriginal people who are able, willing and deemed by the relevant Aboriginal groups as being structurally appropriate to serve on them. Critically, the central role to be played by the GADC in implementing the Agreement and representing the interests of the Native Title Groups could not be undertaken in part because of a lack of the resources necessary to meet its contractual obligations (Pasminco et al. 2002: 94; Scambary 2007: 111–12). While its initial establishment was reasonably resourced, after the first two years CZL was required to provide an annual grant of only $50 000 (indexed) for GADC's operations (Scambary 2007: 112), not even enough to pay the salary of one appropriately qualified staff member, let alone undertaken any of its functions.

The risks posed by the potential non-viability of GADC were identified by GADC itself in the early stages of agreement implementation, but virtually all attempts to seek additional funding, or alternative sources of funds, failed. The author was engaged to work for the three native title groups in setting up the GADC and undertaking its formal functions under the GCA until it had been incorporated and had recruited a staff member to undertake its work. In this role, the author expended considerable effort in seeking to persuade CZL, the

Aboriginal and Torres Strait Islander Commission, various State agencies and others of the implications for the Agreement should GADC not be adequately resourced. Some short-term additional funds were provided by a State agency, but apart from this all efforts proved unsuccessful. CZL in particular argued that meeting any shortfall was the responsibility of the Native Title Groups, and that they should dedicate a portion of the benefit monies received by their eligible corporations to GADC's administration. Not unsurprisingly, this was not supported by the relevant Aboriginal people as they saw these monies as compensation for the damage done to their country, and the impasse could not be resolved. By 2000 GADC was essentially a hollow shell of an organisation with little capacity to undertake its specified functions and little active support from its nominal constituency.

The de facto capacity failure of GADC was compounded by compliance failures in the small Aboriginal corporations receiving the monetary payments from CZL. As previously outlined, one of GADC's functions was to receive an aggregate annual payment from CZL, and hold it in trust pending disbursement to each of six 'eligible' Aboriginal corporations, in accordance with a formula set out in the Agreement. In order to be eligible to receive its payment, each corporation had to maintain compliance with the reporting requirements of its incorporating act, the then *Aboriginal Councils and Associations Act* 1976 (Cth). Within two years, four of the six Aboriginal corporations were ineligible to receive these funds (Martin, Hondros and Scambary 2004: 8) and, of these, three had their memberships drawn from the Waanyi Native Title Group, on whose traditional lands the actual mine was located. Furthermore, recognising the fact that there were many Waanyi people who were not members of or associated with any of the existing four Waanyi corporations, there was an unallocated amount designated for the Waanyi Native Title Group which was received each year by GADC, but which it could not distribute until it had obtained the informed consent of the group as a whole.

The consequence of these entirely predictable compliance failures was that by 2002 a quite substantial sum of money had accrued for the Waanyi Native Title Group which could not legally be distributed by GADC under the terms of the Agreement. However, as explained above, GADC did not have the staff or financial resources to undertake the necessary consultations with Waanyi people about how to resolve this situation, or what to do with the unallocated Waanyi funds. Indeed, it had never had the resources to assist the corporations to maintain regulatory compliance (which would have avoided this problem). The frustration of Waanyi people with the inability of the GADC to pay what they saw as their compensatory monies has been identified as one of the key reasons which led to the 2002 sit-in at the mine site kitchen by Waanyi people, which exposed the operation to serious financial risk at a particularly vulnerable time (Scambary 2007: 236).

This case study provides an unfortunate example of the problems which may arise if agreement implementation structures and processes are not viewed as a system, and agreement governance is not established and implemented systematically. As Fig. 5.1 and the above discussion illustrate, in common with many major mining agreements the GCA constitutes a system comprised of complex, interlinked, and interdependent structures and relationships which are more than just an aggregation of separately negotiated components. However, recognition of this systemic character was not incorporated into its governance or implementation. In such circumstances, an agreement is prone to failure at crucial and unexpected points, and this failure may pose major risks to the interests of *all* parties. It is significant in this context that in the events which led up to the sit-in at the Century mine (which potentially threatened the project), the element of the GCA which failed (monetary payments to the native title groups) was a relatively minor component of the Agreement of little direct concern to the miner or to government. However, these payments were both highly politically symbolic and of intense practical interest to the Aboriginal people concerned.

Governance for transformation

Many (but not all)[4] agreements between Aboriginal people and resource developers are based on procedures set out in the NTA, in particular its Right to Negotiate provisions, precipitated by the successful registration of a claim for native title. To pass the registration test, claimants are obliged to construct an account of their present society and culture in terms of essentially unbroken connections to their pre-sovereignty past—as arising through adaptation to the wider society, not transformation by it. This process becomes even more exacting if claims are to result in a successful determination of native title. From this perspective, the native title claims regime can be seen as a state resourced and mandated project of 'traditionalism'—the reconstruction of an idealised representation of the present as it allegedly is, in terms of how it supposedly was in the past (Merlan 1998: 231). On the other hand, agreements such as Indigenous Land Use Agreements under s.24 of the NTA offer possibilities for Aboriginal people to construct their futures through explicitly transformative processes involving engagement with the institutions of the dominant society.[5] Such processes can enable claimants to negotiate ways to have their interests and certain of their rights recognised and aspirations met (including for development), without these having to be refracted through the distorting lens

---

[4] For example, the Yandicoogina agreement in the Pilbara is a private contract, not an Indigenous Land Use Agreement under the NTA (Scambary 2007: 97, this volume Chapter 8). Also, many agreements have been negotiated in the Northern Territory under the provisions of the *Aboriginal Land Rights (Northern Territory) Act 1976* (Cwlth) (ALRA).

[5] Smith (1998) still offers the most comprehensive account of the Indigenous Land Use Agreement provisions of the Act.

of traditionalism. That is, in contrast to native title claims, agreements are potentially privately resourced[6] and optional projects of modernism.

Native title has been placed in something of a traditionalist policy enclave over the decade since the so called 'Wik' amendments to the NTA. In no small part as a direct result of those amendments, successive Federal and High Court decisions have led to a progressive diminution of what can be recognised as native title, and progressively higher evidentiary thresholds for its proof. As a consequence, native title has become an increasingly attenuated property right, with little direct fungibility to other forms of capital and thus difficult to leverage as an effective base for sustainable development (Pearson and Kostakidis-Lianos 2004, referring to the work of Peruvian development economist de Soto 2000).[7] Native title is also a very legally fragile form of property right. Its existence depends upon continuing adherence by the native title holders to the laws and customs from which their native title derives. Post-determination socio-cultural changes—including indeed those which would logically result from the positive impacts of engagement with the mining industry—could result in a government seeking to have the determination that native title exists revoked, on the basis that the particular groups' laws and customs are no longer traditional.

Paradoxically therefore, while native title (or its assertion) can provide leverage for agreements, its legal fragility provides a poor substrate for agreements in terms of their long-term sustainability. More broadly, an increasing gap has developed between Aboriginal goals and aspirations beyond economic development and what can actually be delivered by the recognition of native title (Strelein 2003). Yet, in the face of overwhelming and continuing disadvantage, under the Howard coalition government Aboriginal affairs policy rhetoric more generally was focused on social and economic engagement, through individual participation in what has been termed (following Pearson 2000a) the 'real economy'—an archetypical modernist project.

There are signs that the Rudd Labor Government may be seeking to move native title out of this enclave, linking it through agreement making into such developing policy frameworks as 'closing the gap' and social inclusion. As discussed at the beginning of this chapter, in February 2008, the Commonwealth Attorney General called for a new approach to resolving native title claims, through 'interests based' negotiations between claimants and other parties resulting in an array of 'non-native title' outcomes (McClelland 2008). More recently, in the 2008 Mabo Lecture the Commonwealth Minister for Families,

---

[6] Although many are also funded by native title representative bodies, most of the large-scale agreements are resourced by the relevant mining company.

[7] While the Right to Negotiate provisions of the NTA, and the veto provisions of the ALRA provide forms of de facto fungibility, it nonetheless is the case that native title rights and interests and those arising from the inalienable freehold issued under the ALRA are not directly fungible to other forms of capital.

Housing, Community Services and Indigenous Affairs argued that benefits from mining agreements should create employment and educational opportunities for individuals and be invested for the long-term benefit of communities. Macklin (2008) observed:

> The challenge here is to ensure that financial flows to native title holders—and indeed landholders under other land rights legislation—contribute positively to improving Indigenous economic status. To do this, these financial transfers must be structured to increase wealth and capital assets within Indigenous communities.

Given that agreements are essentially private contracts between the parties, albeit given certain legal characteristics if negotiated and registered as Indigenous Land Use Agreements, it is difficult to see how government could insist that agreement benefits be structured in certain ways, unless there is the intention to establish Indigenous Land Use Agreements as statutory contracts.

This chapter does not address the question of whether Australian agreements between Aboriginal people and resource developers have historically delivered the benefits which Aboriginal parties expected of them (for example, O'Faircheallaigh 2006),[8] nor the enormous structural and other impediments to their doing so (for example, Taylor 2004a, 2008b; Taylor and Scambary 2005). There are also complex challenges to the policy frameworks of 'closing the gap' and social inclusion posed by the well-documented maintenance of particular Aboriginal worldviews which may be inimical to certain forms of participation in the wider society. Further challenges are posed by evidence that there are many Aboriginal people who, while they seek better access to the goods and services of the wider society, nonetheless have no desire to be assimilated into it or to share all of its values, lifestyles and locales (Altman, Biddle and Hunter 2008; Martin 2005b; Scambary 2007; Sullivan 2007; Sutton 2001).

Whether effectively negotiated and implemented or not, all agreements are explicitly transformative institutions. This is irrespective of whether the aims include the protection and maintenance of Aboriginal culture. An example of this is the GCA, which establishes one of the goals and aspirations of the Aboriginal parties as being:

> ... to ensure that the material benefits do not corrupt indigenous cultures but enable people to re-affirm the cultures and enhance the lifestyles of the members of the Native Title Groups and other members of the Communities through community and cultural development initiatives (GCA 1997: 6).

---

[8] As noted at the beginning of this chapter, there are real issues with the availability of appropriate data for benchmarking and measuring agreement impacts.

Objectives of mining agreements—such as financial benefits and economic development through employment, training and business development—are predicated precisely upon the transformation of the Aboriginal parties' existing socioeconomic status. It is clear in fact that irrespective of the efficacy of mining agreements, or that of government policies, Aboriginal people in mine hinterlands are undergoing profound, and arguably accelerating, transformation. There are significant and ongoing demographic changes in Aboriginal populations in Australian mine hinterlands, as elsewhere in remote and rural Aboriginal Australia. For example, Taylor and Scambary (2005) have produced a major baseline study for Aboriginal participation in the Pilbara mining boom. They showed that in the absence of substantial out-migration the Pilbara Aboriginal population is set to expand for decades to come, with the largest growth being in younger age groups, although there will be a greater proportion of older people than is currently the case. In combination, Taylor and Scambary (2005) argue, these expanding cohorts present significant challenges for social and economic policy. Arguably, they also have major implications for cultural reproduction, with enculturation into a distinctively Aboriginal social and cultural milieu taking place within generational age cohorts—such as peer groups—rather than through transmission from senior to junior generations (Martin 2003, 2008b).

There are other very significant demographic changes taking place. The majority of Aboriginal people associated with any given traditional country are now usually living in polyglot townships along with non-Aboriginal people and members from other Aboriginal groups. These townships may be situated on the country of the group concerned (as Tom Price is for its Eastern Gurrama residents), on or near its periphery (as Aurukun is for its Wik Way residents), or at some remove from it (as is Mornington Island for its Waanyi residents). Those Aboriginal people from a particular country are thus typically dispersed across wide regions, may have only intermittent contact with other members of the group (for example, at funerals and other such ceremonial occasions), and more generally are living in situations where younger generations are exposed to a considerable diversity of values and worldviews. In the terms of French sociologist Pierre Bourdieu (1977: 166) that is, they have moved from a situation of 'doxa' in which the established order had not been perceived as arbitrary and one possible order among many, but as a self-evident and natural one which went unquestioned, to 'heterodoxy' in which there are many possibilities, including those of dissent and rejection.

These factors clearly have major implications for being able to prove native title, in terms of the legal requirements to demonstrate such matters as continuity of the relevant society under whose laws and customs the claimants assert they hold native title, continuing adherence to and practice of those laws and customs through the generations since sovereignty, and the traditional nature of those laws and customs which can allow for adaptation but not transformation. The kinds of changes and transformations outlined above, common across virtually all Aboriginal groups and societies, do not mean that native title cannot be proved or demonstrated, but make it much harder to do so. That is, proving native title requires arguments that overcome these difficulties.

In the negotiation, design and implementation of agreements however, such factors must explicitly be taken into account, not circumvented. For example, there will be a need to develop sophisticated, nuanced, but practical analyses and proposals around such matters as contemporary Aboriginal authority structures and leadership domains. It will be critical to the long-term sustainability of agreements that cultural enclave governance principles such as the supposedly unchanging nature of tradition are not unwittingly built in. Equally, it will be essential to not build in obsolescence such as traditionalist notions of authority of elders in domains where they may demonstrably not have such authority (see discussion below). It will also be vital to be alert to different governance principles which may operate in different arenas—for example, decision-making principles in relation to country, in comparison with those necessary for viable commercial enterprises.

Sustainable agreement governance design will also require detailed attention to be paid to the implications of the complex interplay in Aboriginal societies between the local and individual on the one hand, and the collective or community on the other. There is typically a pervasive public dialogue amongst Aboriginal people around collective social forms, which nonetheless takes place against the background of the reality of an intense localism (discussed later in this chapter) with a stress on local group autonomy, and with ethical and political frameworks centred on highly localised imperatives. One implication of this is that it is likely to be problematic to assume that a small number of representatives from, say, a native title group, will in practical terms prove a sufficient conduit for communication between an agreement entity and the relevant group (see also the discussion below on the politics of representation). As another instance, it may be important in negotiations to not just focus solely on community-based benefits, such as resources and assistance for business development, but also to have the mechanisms for local groupings or families, or even individuals, to access them (see also Holcombe, Chapter 7; Scambary, Chapter 8).

Agreements as intercultural institutions

Thus far, I have discussed the need to incorporate the recognition of agreements as systems, as well as their explicitly transformative character, into agreement governance in the negotiation, design and implementation stages. The final and key element in agreement governance to be considered here is the necessity of reflecting the intercultural nature of the values and practices that the Aboriginal parties to agreements will bring to both their involvement in negotiations and participation in subsequent agreement implementation. However, this is not to propose the apparently straightforward principle that agreements need to take account of Aboriginal culture, for example through creating supposedly culturally appropriate institutions and ways of working within agreements.

This is because while it is possible to meaningfully delineate distinctive characteristics of the contemporary values and practices of the Aboriginal peoples of Australian mine hinterlands, these values and practices have been produced, reproduced and transformed through complex historical processes of engagement with those of the dominant society which has resulted in what Merlan (1998) terms an intercultural social field. This process has involved not just Aboriginal people's domination by and exclusion from non-Aboriginal society, but also their appropriation and incorporation of many of the wider society's values and practices into their own, distinctive, ways of being and acting (Martin 2003, 2005a, 2005b). Even who and what Aboriginal people consider themselves to be has been affected by the representations of Aboriginality by others (Merlan 1998).

Aboriginal societies and cultures cannot be seen as bounded and separate entities or domains (Hinkson and Smith 2005; Merlan 2005); nowhere in Australia do (or indeed can) Aboriginal people live in self-defining and self-reproducing domains of meaning and practices. Rather, they draw from and contribute to complex and contested intercultural social fields (Martin 2003). It should be noted that while the notion of the intercultural implies both Aboriginal and non-Aboriginal people are operating within a (more or less) shared social field, they may well be doing so from quite distinct positions, as Merlan observes (1998: 233). A key insight of significant practical import is the challenge posed by the notion of the intercultural to the existence of separate, disconnected Aboriginal and non-Aboriginal domains of beliefs, understandings, and practices. It recognises that the characteristics of any particular governance arena in, for example, a mining agreement do not draw solely from a supposedly separate Aboriginal domain with its origins in the 'classical' (Sutton 2003) past, nor derive simply from a self-contained and dominant non-Aboriginal society and culture. Rather, each arena or phenomenon will involve values and practices which draw from ideational and practical repertoires whose origins may ultimately lie within either or both Aboriginal society or the wider one, and which are simultaneously implicated in an ongoing cycle of adaptation, incorporation, and transformation (see also Smith and Hunt 2008).

An exemplar of an intercultural phenomenon is that of Aboriginal elders. The contemporary category of elder is not simply a phenomenon of Aboriginal societies themselves with its roots in traditional authority structures; it has been created in part precisely through the interactions between Aboriginal and non-Aboriginal societies (Mantziaris and Martin 2000: 302–3). Elders have become the individuals with whom governments, agencies and resource developers consult to ascertain the views of Aboriginal groups about issues ranging from the protection of heritage and culturally significant sites to the protection of children. In certain contexts, the category has even been introduced into legislation (for example, the *Aboriginal Land Act* 1991 (Qld)). Not only, then, has who constitutes an Aboriginal elder and the nature of their status and authority been transformed by institutions of the wider society, but Aboriginal eldership has in turn impacted on how those institutions interact with and understand Aboriginal groups and communities.

Mining agreements themselves constitute intercultural institutions. They quintessentially arise from and in turn structure and transform the nature of the engagement of Aboriginal people and their institutions with those of the wider society—and vice versa (see, for example, Doohan 2003; Scambary 2007, this volume Chapter 8; Trebeck 2005, 2007a). Indeed, many Australian mining agreements are negotiated on the basis of claimants holding or asserting native title rights in project development areas, and native title itself is an archetypical intercultural phenomenon. As discussed above, its logic derives from the recognition space of native title, rights and interests whose origins lie in traditional laws and customs, but are recognised and given force by the general Australian legal system (Mantziaris and Martin 2000). The intercultural character of native title arises through the processes by which the content of a particular Aboriginal group's or society's relations to country under its own laws and customs is translated into rights and interests which can be accommodated by Australian property law—but which in turn impact on how members of that group or society understand, practice and reproduce those laws and customs (e.g. Glaskin 2007; Redmond 2007).

While a mining agreement as a whole needs to be understood as intercultural, so too do its constituent entities and relationships and, more broadly, the economic values and motivations which Aboriginal people bring to bear on their engagement with a given mine and its associated agreement. It is not just specifically Aboriginal institutions such as the GADC previously discussed in relation to the GCA which are intercultural. Equally, key functional areas of the mining company or its subcontractors such as a Community Relations division charged with the responsibility for engaging with Aboriginal stakeholders and oversight of agreement implementation are also potentially intercultural institutions. Their organisational cultures (such as work practices, accountability constituencies, styles of interpersonal relationships and more generally an ethos which are all influenced by the values and practices of Aboriginal staff and clients) have the potential to be quite different from those of, for example, production units, as is the case with the mine site GCA Support Department within CZL (Scambary 2007: 224–5; Trebeck 2005). On the other hand, Rio Tinto Iron Ore has taken a different course with its Communities and External Relations division, on the basis that to be relevant it must share common corporate goals and culture, with community relations being an integral part of the company's overall operating strategy (B. Hart, pers. comm. 2008).

In these core project operational areas (such as the mine itself, crushing and beneficiation plants, and assay laboratories) factors such as production targets, quality control, other technical requirements, and occupational health and safety standards, limit most potentially transformative impacts of Aboriginal involvement on what is overwhelmingly modern industrial production culture. On the other hand, mine administrative and service operations may be more open to other cultural influences. In the Century project, factors such as the significant presence of Aboriginal employees (in the range of 15–20 per cent according to Barker and Brereton 2004), political actions such as the 2002 sit-in previously discussed, and proactive leadership by both Aboriginal and non-Aboriginal individuals have served to create a distinctive culture around the mess and recreation and living areas which marks this project out from other field sites discussed in this volume. In the case of the Argyle diamond mine, however, Doohan (2003) provides examples of how the Argyle Participation Agreement between the miner and traditional owners has allowed the latter to 'insert their cultural forms and presence onto the mine site in a number of ways' (see also Aboriginal and Torres Strait Islander Social Justice Commissioner 2007, Chapter 5). These have included Aboriginal people framing their relationships with the miner in terms of *wirnan* exchange relationships, instituting regular *manthe* ceremonies involving the (Aboriginal) hosts welcoming and inducting (company staff) guests and giving them a ritual safe passage across the mine site—which in Doohan's view constitutes a form of a specifically Aboriginal health and safety instruction, and social incorporation of senior mine personnel through giving them 'skin' classificatory names.

Returning to the implications for agreement negotiation, design and implementation, while a range of entities and relationships within a mining agreement need to be understood as intercultural, the intrinsic character and the entailments of interculturality of each will be potentially different. That is, while the governance of entities such as an Aboriginal representative and advocacy body, or the company's community relations and employment divisions, and relationships between each of these entities and the Aboriginal stakeholders must all incorporate a recognition of intercultural factors, different issues will arise for each in their design and implementation. The following factors are relevant to defining the intercultural character of each arena. These are crucial matters to be established for both design and implementation purposes:

- Is the governance that of a relationship between an entity and a collectivity, between two entities, of relationships within a collectivity, or of an entity itself?

- Does the particular governance arena entail multiplex linkages or is it relatively mono-dimensional?
- Does the institution involved have a formal, legal or administrative presence, or is it a collectivity or a 'natural social grouping' of some kind?
- What is the source of authority for the relevant principles of governance (for example, to adjudicate on conflicting viewpoints, resolve disputes, and to establish the rules of practice)?

A schematic illustration of entities and the relationships between them to be found in Australian mining agreements with Aboriginal people is provided in Fig. 5.2. It is not intended to represent the formal structures of any given agreement as Fig. 5.1 does in the case of the GCA. Rather, it illustrates governance *arenas* relating to classes of entity and categories of relationships in such agreements.[9] Its aim is to disaggregate different kinds of governance arenas in order to illustrate how the preceding proposed governance principles—the need for agreements to be understood as systems, agreements' transformative character, and their intercultural nature—can be usefully brought to bear in specific instances. That is, it aims to break down agreement governance into components which potentially need to have their own distinctive and specific governance characteristics to support agreement sustainability.

Fig. 5.2 Key agreement governance arenas

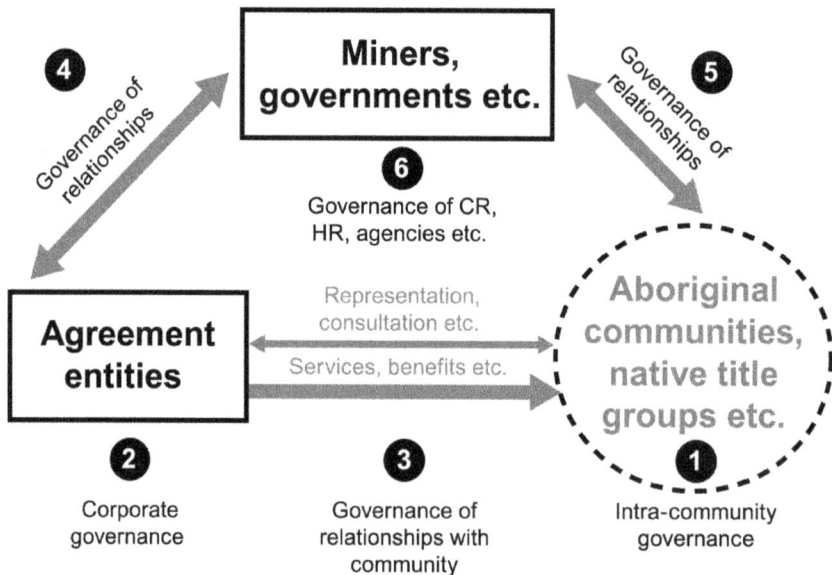

---

[9] In Fig. 5.2, the line representing Aboriginal groups and communities is dotted to indicate their relative lack of clearly defined bounds in comparison with, for example, the formal membership of a corporation.

Because the focus of this chapter is neither on the governance of mining company community relations divisions and the like, nor on the governance of governments in terms of how its various agencies may operate in their dealings with Aboriginal people through mining agreements (see O'Faircheallaigh 2006 for some discussion of these matters; also Trebeck 2007a, this volume Chapter 6), these have been aggregated in Fig. 5.2. Effective governance in each of the six arenas identified in Fig. 5.2, I propose, is one of the key requirements for the long-term effectiveness and sustainability of agreements as a whole. Each arena will have differing intercultural characteristics, and different transformative implications—and thus potentially involve different governance principles. For illustrative purposes however this chapter will sketch these out in only two of these arenas of particular significance to mining agreements: first, the governance of agreement organisations themselves *qua* organisations—that is, corporate governance and the like; and second, the governance of relationships between agreement entities and the particular Aboriginal stakeholders.

The governance of agreement structures

This section focuses on Arena 2 in Fig. 5.2, concerning the governance of more or less formal entities such as committees and working groups as well as incorporated bodies, although discussion here will centre on the latter. There are now several thousand Aboriginal-controlled organisations around Australia (Hunt and Smith 2006a: 10), ranging from virtual 'post box' landholding entities to commercial and service delivery corporations with turnovers of millions of dollars. As is the case for non-Aboriginal organisations, there is also considerable diversity in terms of their viability or otherwise. There is a developing, if contested, literature on theorising Australian Aboriginal organisations and governance (for example, Mantziaris and Martin 2000; Martin 2003, 2005a, 2005b; Martin and Finlayson 1996; Sullivan 2006, 2007), and the factors that are held to contribute to good governance and successful Aboriginal organisations (for example, Finlayson 2007a, 2007b; Hunt and Smith 2006a, 2007; Hunt et al. 2008).

However, this section will not canvass this terrain in any detail, but will refer to a set of key issues that bear on the design of organisations such as those established through mining agreements. In particular, it will follow arguments that it is necessary to separate—both conceptually and practically—the governance of organisations themselves (Arena 2), from that of the communities which they serve (Arena 1), and that of the relationships between them (Arena 3) (see, for example, Mantziaris and Martin 2000: 126–8; also Martin 2003; Sullivan 2007).

Of course, Aboriginal people bring distinct values and ways of acting in the world to their participation in organisations, which have become fundamental constitutive elements in Aboriginal polities. Some of these pose challenges to the governance of Aboriginal-controlled organisations, especially those with diverse and broad constituencies. An 'intense localism' (Martin and Finlayson 1996) is a particularly significant feature to be found across Australian Aboriginal societies generally, with ancient roots in their original hunter-gatherer predecessors. Here, priority is given to values and interests asserted at the small-scale, locally based or even individual levels, and to individual and local-group autonomy (Martin and Finlayson 1996: 5; Sullivan 2006: 17). While this localism exists along with wider networks of connection and interdependence (Myers 1986; Hunt and Smith 2006a: 24–5), strong emphasis is typically placed on the identities and autonomy of individuals and local groupings, such as those referred to by Aboriginal people as 'families' which frequently form the 'backbone' of organisational governance (Hunt and Smith 2006a: 10).[10]

Localism has important ethical as well as political implications. A person's strongest bonds and obligations are usually to their immediate kin and family, and those from other groups may well be viewed with a degree of suspicion or even hostility. A notion of the wider common good—including amongst those who are parties to the one mining agreement—may not meaningfully exist, or be very attenuated (Peterson 2005; Trigger 2005; Tonkinson 2007). These overriding political and ethical commitments to immediate kin and family have significant implications for Aboriginal members of governing committees and boards. To take just one instance, the legal requirement for a board member to act in accordance with their fiduciary duty to the organisation itself can directly conflict with the ethical and political requirement that the individual concerned act in the interests of and support immediate kin, including directing organisational resources to them (for example, Hunt and Smith 2006a: 17; Mantziaris and Martin 2000: 189–92).

Equally, principles underlying democratic representative institutions and other organisational structures whereby individuals or groups cede their right to speak for, manage, and protect their interests to others who represent them, do not typically operate within Aboriginal groups and communities. The localism mentioned previously is manifested in resistance by individuals and local groups to others taking on this role—and indeed to being bound by the decisions of others, including those who nominally act for or represent them in contemporary institutions. This originates in part in the high value placed on individual and local-group autonomy, and on a resistance to hierarchy outside the religious and ritual arena. Equivalently, a person's occupation of a formal institutional position such as chairman or board member does not necessarily give that person the authority or legitimacy within the relevant Aboriginal polity to speak for others.

---

[10] Sutton (2003, Chapter 8) provides an extended treatment of post-classical Aboriginal 'families of polity' as fundamental political, social and economic forms in contemporary, 'post-classical' Aboriginal societies.

All Aboriginal participants understand and acknowledge this cultural logic and imperative by which the primary ethical and political obligations of those on a board are to *their* kin and mob—and that therefore those from outside that network can have no necessary assurance that their interests will be fairly and equitably represented.

This is one of the key reasons why, across Australia, many Aboriginal organisations are characterised by intense competition between different groups and by corporate histories in which competing factions alternate in their control of the board, or fission off to form new organisations. Political conflict in these organisations will often be conducted through manipulation of membership and meeting processes to establish control of boards (and therefore of organisational resources). This destabilising 'politics of representation' (Mantziaris and Martin 2000: 303–5) can also be seen sometimes in the attempts by individuals and sub-groups within the wider organisational constituency to assert control over the means by which membership of the organisation, and thus the means by which composition of its board, are determined.

Nonetheless, while distinctive, Aboriginal organisations are not cultural isolates but focal sites where Aboriginal practices and values are both incorporated and simultaneously transformed through processes of engagement—appraisal, contestation, and appropriation—with those whose ultimate origins lie in the broader non-Aboriginal society (Martin 2003: 5). Aboriginal organisations have become sources of legitimacy and authority not only within the Aboriginal domain but also the non-Aboriginal one where, in some respects, they can be seen as the functional equivalent of the King Plates the early colonists were wont to hang around the necks of putative leaders of local Aboriginal groups (Mantziaris and Martin 2000: 101, 274). They have a form of dual incorporation, whereby they are simultaneously legally incorporated under, or established by, statutes of the general Australian law, and incorporated into Aboriginal polities (Mantziaris and Martin 2000). Aboriginal organisations are thus necessarily and intrinsically intercultural institutions (Martin 2003)—not culturally autonomous Aboriginal arenas, but rather the locations of transforming and transformed practices and values.

From this perspective, then, I suggest it is totally inadequate to leave the construction and evaluation of organisational management principles solely to the Aboriginal people concerned and to a domain of supposedly uniquely Aboriginal values (Martin 2003: 9). If good organisational governance is a core component of an increased capacity by Aboriginal people for strategic engagement with the dominant society (Martin 2003), then it must draw not only from the values and practices of Aboriginal people, but also from those of the general Australian society. While the possibility of distinctive values and practices must be accepted as a basic premise in institutional design, the essence of developing appropriate Aboriginal organisational governance does not lie in supposedly resolving potentially conflicting cultural values and practices; rather, it is to be undertaken through establishing institutional structures and principles which are robust enough to encompass and engage diversity, competition and conflict, and which are appropriate to the task at hand.

It is therefore not defensible to resort to an unexamined notion of cultural appropriateness, or to one of a notionally autonomous domain of Aboriginal culture, in determining the core principles by which effective Aboriginal organisations should be established and operated (Mantziaris and Martin 2000: 293–4; Martin 2003: 9–10; Sullivan 2006, 2007).[11] The concept of cultural appropriateness in relation to Aboriginal organisations assumes a domain in which Aboriginal values and practices are autonomous from those of the general Australian society, and a domain of operations of these corporations which is separate from the legal, political, and economic fields in which they are necessarily situated. Neither assumption is true. As I have argued elsewhere:

> The more attempts are made to reflect the complexities and subtleties of the values and practices of Indigenous people in formal corporate structures and processes—for example, regarding such matters as authority and decision-making, or the various forms of the typically labile Indigenous groupings and sub-groupings—the more there is the risk that the formal corporate structures and processes over time will supplant the informal Indigenous ones—a process of the juridification of social relations. While as we have seen, the engagement of Aboriginal and non-Aboriginal people can best be understood in intercultural terms, juridification takes this a step further, raising the problem of the underlying social relations being distorted or dominated by the legally enforceable expression of the same relations (Martin 2003: 10).

A corollary is that attempts to import particular aspects of Aboriginal political culture into the management structures and procedures of an organisation run the risk of creating organisational instability, as in the phenomenon of the politics of representation mentioned previously. This can have highly adverse consequences for Aboriginal people themselves, as in cases where the organisation is delivering an essential service, or managing a multi-million dollar trust. The objective fact is that representative structures can never truly reflect the nature of and relationship between the fluid and diverse groupings and alliances that characterise Aboriginal political systems. It is a common mistake, repeated across Aboriginal Australia including by those providing advice on the establishment of Aboriginal corporations, to focus attention largely on attempting to capture

---

[11] Sullivan (2006, 2007) challenges the related notion of 'cultural match' developed by the Harvard Project on Native American Economic Development and adopted in the work of CAEPR's Indigenous Community Governance Project.

the diversity of the particular constituency in the board structure itself. It can certainly be important to an organisation's legitimacy with its constituency that its board is representative, in the particular sense of being drawn from a broad cross-section of the constituency and reflecting as far as feasible the cultural geography of the governance environment (Hunt and Smith 2006a: 24). However, problems will inevitably arise when the work of incorporating and responding to the diverse interests and expectations across an organisation's Aboriginal constituency is left essentially to the political and administrative representative structure of a board.

From this perspective, there are compelling arguments for establishing Aboriginal organisations which leave as much distinctively Aboriginal social and political process as possible within the informal Aboriginal realm where it belongs, and do not attempt to codify it within corporate structures or organisational governance mechanisms (Martin 2003: 10). The focus in these corporations' design and management should be on instituting organisational governance of a form which will maintain a viable and legitimate structure through which services would be delivered. Mantziaris and Martin (2000: 322–7) outline a set of principles for organisational design, which while developed for Prescribed Bodies Corporate are of more general applicability. These are: legal certainty; legitimacy (in the sense of having the capacity to attract the allegiance of the group); sensitivity to Aboriginal values; sensitivity to motivational complexity; revisability; robustness; simplicity; and transactional cost efficiency.

As Mantziaris and Martin note, there are interrelationships between these principles, and tradeoffs between them; for example, between sensitivity to Aboriginal values, and transactional cost efficiency.

Separate attention needs to be paid to such matters as developing procedures to ensure effective and accountable relationships and linkages between the corporation and the relevant Aboriginal group or community (Arena 3, Fig. 5.2; see discussion in the following section). These are the areas where full cognisance must be taken of the informal and pervasive governance principles operating within the relevant Aboriginal community itself. In Sullivan's (2007: 15–6) terms:

> A developmental or service delivery organisation should not be conflated with an institution of self-government. It needs neither a representative structure nor should it attempt to mimic local cultural forms. The representative structure is not required because the function of representation continues to happen where it belongs, in the cultural milieu of the community, and in the forms appropriate to the culture. Attention should turn away from representative structure (in service-delivery organisations) and towards means of communication, information transfer (in both directions), monitoring of consent, and effective policy input from the client/membership/constituency. This

means seeking authority wherever it lies, whether in institutions, families or particular individuals and encouraging sound leadership.[12]

This is not to argue that representative Aboriginal organisations are not necessary to mining agreements—they are. But, it is to propose a reconceptualisation of their character, purpose, and therefore design. The challenge is to develop and manage distinctively Aboriginal organisations which nonetheless facilitate effective engagement with the wider society rather than limiting it (Martin 2003: 10), and which pay particular attention to the governance of their relationships with their constituencies. From this perspective, appropriate and effective Aboriginal organisations would not draw their structures and operating principles from a supposedly autonomous Aboriginal domain, but from universal standards of good management (Sullivan 2007: 15–16). Indeed, the scale of many Australian mining agreements is such that there is an overriding necessity for highly competent management. While Aboriginal organisations must certainly take account of specific values and practices of the Aboriginal people who participate in them and whom they serve through the development of flexible organisational cultures which are sensitive to the milieu in which they operate, to be truly culturally appropriate Aboriginal organisations will also have to directly engage—and even on occasion challenge and circumvent—these values and practices. Where it is absolutely essential that Aboriginal values and practices be taken into account is in how they do their business; that is, in the governance of their relationships with their Aboriginal constituents and service community. Aboriginal governance needs to move beyond concentrating on the structure of organisations and towards the development of effective consultation, information sharing, and permission mechanisms (Sullivan 2006: 18, 2007: 30).

Governance of relationships with Aboriginal stakeholders

My focus in this section is on Arena 3 in Fig. 5.2, the governance of relationships between (Aboriginal controlled) agreement entities and their stakeholders. Before proceeding to discuss this, it must be noted that major mining agreements also include another and very important class of relationships involving the Aboriginal stakeholders, that between them on the one hand and the resource developer and relevant government agencies on the other (Arena 5, Fig. 5.2). A key focus of this arena is typically on such matters as education, training, and employment programs, which are beyond the scope of this chapter (see, for example, discussions in Barker and Brereton 2004; Scambary 2007: 86–9, 224–6; Tiplady and Barclay 2007; Vidler 2007). It is clear that to be effective, service delivery in this area must proactively engage potential Aboriginal participants through culturally aware, flexible measures adapted for their specific needs and

---

[12] In my reading, Sullivan is here using 'representative' in the sense of political representation, not the sense discussed previously of a 'representative sample' of constituents.

circumstances, as instanced by the recruitment case studies for the Argyle diamond mine in Tiplady and Barclay (2007: 24–39). Similarly, Kemp, Boele and Brereton (2006) argue for a more proactive, externally focused, stakeholder-driven and values-based approach by resource companies to community relations, another aspect of governance of relationships (Arena 5).

On the other hand, the need for proactive, flexible and adaptive mechanisms for engaging with Aboriginal people does not, in general, appear to have been a widely instituted principle in the relationships between Aboriginal stakeholders and the Aboriginal-controlled agreement structures such as trust companies and special-purpose representative and advocacy bodies (that is, community relationships in Arena 3, Fig. 5.2). As has been argued above, typical agreement governance arrangements have left the work of reflecting community diversity largely to formal structures like boards and committees, like those in the GCA outlined previously. Furthermore, the common expectation of miners and governments is that community members on these structures will act as effective conduits of information, concerns and issues between Aboriginal people and other parties to agreements. Implicitly, the assumption is that representative boards and committees can act as proxies for the relevant community. Direct engagement with Aboriginal stakeholders, it is assumed, is to be conducted largely through community meetings of one sort or the other. Neither assumption can be safely made.

The first assumption—that community representatives on boards and committees will necessarily act as effective conduits for information—fails to take account of the import of localism in establishing people's social, political and ethical frameworks. Even though a nominated or elected member of a board or committee may notionally represent a particular sub-group of the agreement beneficiaries (for example, a language group), communication may not necessarily flow across boundaries set by immediate kin and family connections—much depends on the individual concerned.

The second assumption—that meetings provide a key mechanism for communication with Aboriginal people—also needs careful examination (see also Sullivan 2006: 29-30). Meetings provide a problematic mechanism for informing and seeking input from members of what are often dispersed and deeply factionalised groups. This is for a number of interrelated reasons, to be found across Aboriginal Australia and which in part reflect the highly localised nature of Aboriginal polities discussed previously. Particularly in the case of large, community meetings involving people from disparate groups, they are prone to being dominated and disrupted by individuals who use them for political aggrandisement, which can mean that it is difficult if not impossible to ensure effective participation of those who may have equivalent rights but less political standing. Such meetings can also provide forums that become dominated by the airing of grievances about the operation of the agreement as it is the only occasion available to do so.

Meetings usually provide a poor basis for informed decision-making, particularly around complex and technical issues. Meetings do not facilitate typical Aboriginal decision-making processes of extended consideration and discussion, involvement of appropriate individuals on the basis of such principles as seniority and legitimate knowledge, and consensus building within the local groups where such processes have force. Indeed, because large meetings aggregate local and autonomous groups, they can disempower many people from effective participation. Meetings are a useful and necessary mechanism by which formal ratification of a proposal can be given by the relevant jural public (Sutton 2003)—that is, by those who can legitimately express a position on it and those others who act as witnesses to the ratification. These meetings need to be preceded by a considered and dispersed process of information dissemination, consultation and consensus building, so that the meeting essentially ratifies informed decisions that have already been reached, within the appropriate-level subgroups such as families.

The common reliance upon representative structures together with consultations and information dissemination through meetings as primary means of agreement beneficiary involvement, can also lead to the development of an inherently passive and reactive relationship between beneficiaries and the agreement entities which provide benefits to them. In general, in Australian agreements there appears to be little or no provision for planning or decision-making in which the beneficiaries themselves are or can be actively involved. As a consequence, the dominant relationship between beneficiaries and agreements has become in a number of instances one of opportunistic rent seeking by the former. This is because there are few formalised means by which the beneficiaries can access benefits or be involved in agreement operations or its decision-making processes, and so they are reduced to instrumentally seeking individual or family advantage. Assiduous demanding of sitting fees for mere attendance at meetings is but one example of such understandable, but arguably problematic, rent seeking.

Furthermore, the socioeconomic profiles of Aboriginal people in Australian mine hinterlands, evidence very poor health, education level, employment histories, and so forth (for example, Taylor 2004a, 2008b; Taylor and Scambary 2005). This is a poor substrate on which to graft the important work of agreements in such areas as economic development, human capital development, and other objectives to be found in Australian agreements. Capacity development amongst the beneficiaries therefore is an absolutely essential precursor maximising the returns of the resources provided to leverage sustainable change in accordance with agreement objectives. To be effective such capacity development must operate at the level where beneficiaries themselves also operate—primarily at the individual and local group levels.

These factors thus suggest that the governance of Arena 3, the relationship between agreement entities and the beneficiaries, should have as its guiding principles:

- providing mechanisms for active participation amongst the beneficiaries at individual and local group levels;
- replacing current reactive and passive relationships between most beneficiaries and agreement entities with relationships based on active participation and a sense of ownership;
- minimising opportunistic rent seeking by agreeing on structured processes in which beneficiaries have a meaningful say in the operations of agreements, while still maintaining appropriate mechanisms for prudential control;
- providing mechanisms (such as participatory cyclical planning processes) by which beneficiaries can plan for their futures and the role which resources from agreements might play in those futures at the levels which are meaningful to them (such as family or residential group) to develop a long-term development perspective in the agreement; and
- a pre-eminent focus on working with the beneficiaries to build their capacity to undertake this planning.

Conclusion

This chapter has argued that the governance of agreements between resource developers and Aboriginal people is a crucial aspect of negotiations, and a critical implementation issue. This suggests that governance capacity (for both Aboriginal and non-Aboriginal parties to an agreement) should be developed as much as possible well ahead of agreement implementation. Governance needs to be developed and implemented with agreements considered as systems rather than just as aggregates of disconnected entities, relationships and processes. Agreement governance should be understood as intercultural, not as involving interaction between discrete and disconnected Aboriginal and non-Aboriginal cultures: it should not be designed and implemented in cultural enclave terms.

Furthermore, there is a range of governance arenas in agreements, each of which will exhibit different characteristics and require different governance principles for sustainable agreement implementation. In particular, it has been argued, the focus historically has tended to be largely on incorporating Aboriginal diversity and distinctive values into the structures of Aboriginal organisations, while insufficient attention has been paid to the governance of the relationships between these entities and their constituents, clients, or beneficiaries. Finally, and critically, the negotiation, design and implementation of agreements should explicitly take into account the profound transformative processes involving

Aboriginal engagement with the institutions of the dominant society and not implicitly be predicated on idealised representations of Aboriginal society and culture.

6. Corporate responsibility and social sustainability: Is there any connection?

Katherine Trebeck

This chapter examines the relationship between corporate social responsibility and social sustainability on the premise that a vital ingredient in social sustainability is the capacity of communities to determine, or at least influence, those decisions that impact them. Communities should be able to influence decisions regarding any trade-offs that may affect them—for example, between economic development and environmental conservation or between meeting the needs of current generations and the capacity of future generations to meet their needs. Local communities[1] in particular, need to determine what is to be sustained, how and at what expense—not only because their lives are impacted by these decisions, but also because local communities are crucial players in processes seeking to enhance sustainability. Corporate actions are clearly relevant in these trade-offs, but if, as seen here, the state proves itself a poor mediator between company behaviour and the desires of local communities, then local communities need to confront companies directly.

It is this interface that is the subject of this chapter—how responsive companies are to the demands of local communities and in what circumstances. It considers what drives miners to respond to the demands of certain communities in order to elucidate when and where communities, in all their complexity, can exert leverage over companies to make corporate behaviours correspond more with the desires of those affected by corporate actions. The case study of mining companies and their relations with Indigenous communities in Australia is used to evaluate whether the tool of 'civil regulation' might advance the sovereignty of local communities, so contributing to sustainability and enhancing democracy.

Four examples indicate that corporate responsiveness depends on individuals inside companies perceiving and utilising the 'business case' for socially responsible behaviour to pursue corporate change in the direction demanded by particular communities. This 'business case', however, depends on communities maintaining vigilance and sustaining a context that impels companies to respond to community demands. Implications for social sustainability arising from this will be set out in the conclusion. Before examining

---

[1] Local communities can be seen as linked by a shared space—thus 'communities of geography'—other communities might be linked by a mutual concern in an issue—'communities of interest', or by common experience of being affected by a particular development or entity—hence, 'communities of fate' (see Held 2004, who describes the world as comprising 'overlapping communities of fate'; also Hirst 1994: 49).

several 'vignettes' where civil regulation has been effective theories of corporate social responsibility, democracy and social sustainability, and civil regulation are outlined to frame discussion. Instances of civil regulation and corporate change are the Century mine negotiations and sit-in, Hamersley Iron's Marandoo dispute, Rio Tinto's adoption of corporate social responsibility, and the campaign to stop the Jabiluka uranium mine.

Analysis is informed by over 120 semi-structured interviews conducted from 2002 to 2005 with all levels of company personnel and many industry observers, regulators, bureaucrats, stock market participants, environmental and community activists and academics, as well as consideration of corporate publications and other quantitative sources to contextualise personal accounts. This supporting material included annual reports, corporate statements of a company's vision and operating principles, websites, brochures and other publications, survey questions and results, speeches, media releases, media reports and company statistics.

Democracy and social sustainability

Democratisation is advanced when political decision-making becomes more inclusive—when citizens become involved in those decisions that affect them (Whitehead 2002).[2] The capacity of communities to determine, or at least influence, those decisions that impact them is a fundamental aspect of sustainability. Advocates of more participatory modes of democracy, however, often complain that contemporary configurations of parliamentary representation are inadequate (see, for example, Barber 1984; de Tocqueville 1945; Dryzek 2000b; Fox and Miller 1995; Hirst 1994; Odd Var Eriksen 2000; Pateman 1970; see also Rayner 1997 for a comprehensive account of the pit-falls of Australia's representative structure). Formal democratic political processes, invariably parliamentary representation, are accessible to some groups more than others, reinforcing social and economic disparities, and deepening political inequality (Young 2000; see also Dahl 1985; Klausen and Sweeting 2003). Globalisation and corporate power seem to be encroaching on the ability of representative government to uphold citizens' interests (see, for example, Bang and Bech Dyrbery 2000; Etzioni-Halevy 2003; Korten 2001; Miller 2003). Consequently, government action frequently does not reflect the wishes of particular communities.

With access and influence in formal parliamentary democracy often determined by possession of resources, Indigenous Australians are especially disenfranchised by the contours of representative democracy. A diversity of factors—historical, political, cultural and structural—can potentially explain why many Indigenous communities are disadvantaged in the socioeconomic conditions they experience

---

[2] The etymological meaning of 'democracy' is rule by the people.

compared to non-Indigenous Australians (see, for example, Altman 2001b). For example, Indigenous people are more dispersed than other Australians, many live in remote regions with few employment opportunities, while poor education, housing, health and income status reinforces their disadvantage. Consequently, Indigenous Australians are significantly more likely to be impoverished than non-Indigenous Australians—across many indicators.[3] Compared to non-Indigenous Australians, unemployment is 3.2 times higher for Indigenous peoples and Indigenous life expectancy is 17 years lower than that for the total Australian population (Steering Committee for the Review of Commonwealth Service Provision 2005).

Such circumstances illustrate that while Indigenous people are, in theory, able to access representative parliamentary structures on the same basis as non-Indigenous Australians, their standard of living or outcomes from education or health services are substantially below that of non-Indigenous Australians (Robinson and Sidoti 2000; Stokes 2002; Westbury 2003). Despite the complex of reasons for Indigenous disadvantage, imposed 'solutions' developed in distant centralised government clearly do not meet the needs of many Indigenous communities.

Furthermore, until relatively recently, Indigenous Australians enjoyed few formal legal and political rights, and struggled to access many mainstream government services. Limitations of parliamentary representation are starkly evident in the Indigenous experience: the small number of Indigenous voters and their geographic dispersal translates to little electoral 'muscle'. In a democratic system largely dominated by formal parliamentary representative politics, many Indigenous communities have been marginalised to the extent that they are unable to acquire political influence necessary to meet their objectives. In addition, disenchantment with formal structures and processes of representative parliamentary democracy has been deepened by concerted government efforts to facilitate mining, often regardless of the articulated opposition or concerns of local Indigenous communities (seen in the examples below). Such government actions over-ride the wishes of local communities, undermining the influence of Indigenous citizens in decisions that acutely affect them.

In addition, substitution—the risk of citizenship entitlements being deliberately reduced—is faced by some Indigenous communities where mining takes place. Government substituting royalty or other benefits derived from mining in the place of government provision results in a reduction of state finance for community services (Altman, Chapter 2; Levitus, Chapter 4; Altman and Pollack 1998; Banerjee 2001; O'Faircheallaigh 2004a; Rowse 2002).[4] This is made more

---

[3] Scale of measurement used does not change this fact (Hunter 1999).
[4] For suggestions on how best to avoid this situation see O'Faircheallaigh 2004a.

complex from a democratic perspective: perceptions that existing lack of government service delivery (or threatened reduction) might lead some Indigenous communities to acquiesce to mining if they view mining as the only means to obtain necessary outcomes—health and education services or employment, for example.

The deficiencies of representative democracy thus necessitate supplementary means of delivering democratic ideals such as self-determination and inclusion of those affected in policy-making (Hirst 1994; see also Dryzek 1996b, 2000a; Fliedner 2000; Goghill 2002; Rose 2002). Legitimacy and democracy should not be considered simply as a function of the tenure of elected officers, but should be judged according to how institutions (including non-state entities) address the expectations of those communities they impact. This broad understanding of democracy and governance is further warranted by apparent ability of certain groups to attain desired behaviour from powerful socioeconomic entities—including companies (see, for example, Dryzek 1996a; Eckersley 2000: 118; Maddox 1991; Moon 2003: 2–9; Pateman 1970; Suggett 2000).

If Indigenous communities can, using the tool of civil regulation (see Trebeck 2007a), compel companies to reflect their demands, then the sovereignty, self-determination and sustainability of these communities has been advanced. How this tool is manoeuvred to attain community expectations depends on several factors, pertaining to both the communities and the company in question. The examples below seek to elicit some of these factors to enhance understanding of the connection between corporate social responsibility and social sustainability.

Corporate social responsibility: A definition

This chapter conceives corporate social responsibility as those company activities, other than commercial outputs, and beyond legally required behaviour, taken in order to satisfy social needs and demands (see Trebeck 2008a). Motivation is of particular importance in this conception: these are activities taken in response to community demands or with a view to addressing social needs that impact the business. They are ultimately impelled by business needs, as opposed to moral or ethical connotations sometimes implied by 'responsibility'. The pragmatic rationale of such responsiveness reflects that in order to serve shareholder interests, attention to the demands of some groups beyond shareholders is necessary (see, for example, Moon 2003: 2–9; Power 2003; Spar 1998). According to a recent report, this might be understood as a business approach to social responsibility, seeking the company's own self interest to enhance long term value and risk reduction[5] by addressing the social and

---

[5] Non-financial risks might include health and safety risks, protecting physical assets, compliance with regulations, product liability, brand reputation, asset vulnerability, changing markets, sabotage, human capital and so on (see Corporations and Markets Advisory Committee, 2006: 50).

environmental operating context (a product of factors such as reputation, employee attraction, improving market position) (see Corporations and Markets Advisory Committee 2006: 34–68). Alternative approaches to a company's social responsibilities include a compliance approach, a philanthropic approach, and social primacy or social obligation approach. The compliance approach is based on obligation to comply with the letter of the law; at the same time, companies may benefit from complying with the spirit of the law as perceived by the general community. Under a philanthropic approach companies give to the community beyond their primary business. Under a social primacy approach directors pursue ethical goals beyond the spirit of the law, often in recognition that businesses have access to valuable resources and privileges, and therefore have a corresponding obligation to address social problems, even if there is not a clear benefit to the company (Corporations and Markets Advisory Committee 2006: 52–3).

Civil regulation

Bendell defines civil regulations as those 'pressures exerted by processes in civil society to persuade, or even compel, organisations to act differently in relation to social and environmental concerns' (Bendell 2000a; see also Murphy and Bendell 1999).[6] A particular manifestation of civil regulation with salience for mining companies is the necessity to obtain and maintain a 'social licence to operate'. While minimum levels of acceptable company behaviour are specified in laws and regulations, social obligations are no longer discharged by simply carrying out legal minimum duties. Companies need to be not only 'morally and legally acceptable, but also popularly acceptable' (Fombrun 1997; Holme 1999; Lane 2001; see also Gunningham, Kagan and Thornton 2002; Piggot 2002; Warren 1999: 214–24; White 1999: 30–43). To attain this acceptability companies must be deemed to meet community expectations for certain behaviour, often beyond behaviour mandated by formal state regulations. In seeking a 'social licence to operate', corporate responsiveness to the demands of key communities brings companies to act, to some extent, according to the demands of certain communities (Moon 2003: 2–9).

Gunningham, Kagan and Thornton (2002) highlight that the terms of a company's social licence can be indirectly stipulated via mechanisms offered in the law and the economy: for example, through unofficial economic sanctions associated with negative publicity that reduces sales, investment in the company, and the company's access to capital (see also Kapelus 2002: 275–96).[7] As seen in the case

---

[6] Corporate power has long been challenged by elements of society and social movements, such as the consumer movement in the 1960s, and the environmental movement of the 1970s. The history of business/civil society relations has been one of antagonism, until the early 1990s when partnerships for sustainable development began to emerge (see for further description Murphy and Bendell 1999).
[7] Economic enforcement mechanisms invariably entail boycotts such as by consumers, investors or shareholders in their capacity as economic actors.

studies below, local communities can also directly affect the commercial future of a company by threatening its ability to attain resources, such as employees, minerals, or access to operations using blockades, legal permissions or licences.

Community resistance can also be tacit, such as intransigence during land access negotiations. Or community resistance can be more palpable, such as public protest, physical occupation of a site, even sabotage. Direct action can in turn generate wider, indirect, implications. For example, causing a project to be delayed lowers the net present value of any future earnings steam.[8] Moreover, local communities can use particular legislation strategically to exert leverage over companies. Similarly, formal regulations might be more rigorously enforced, or more stringent regulations introduced, as the result of community pressure.

Many companies are consequently recognising a 'pragmatic logic' in meeting community expectations given ability of local communities to impose costs on operations (Kapelus 2002: 275–96). Good corporate community relations, community engagement and efforts to meet community demands are the means by which companies can improve reputation and attain acceptance: their social licence to operate (Banerjee 2001; Cragg and Greenbaum 2002: 319–35; Parker 2002; Solomon 2000; Trebeck 2007a; Zadek 2003).

Social sustainability through civil regulation and corporate social responsibility?

In order to explore the bearing these processes have on social sustainability in the sense of community self-determination, several empirical examples are outlined here to assess whether the vulnerability of companies to civil regulation actually empowers those communities otherwise disempowered by formal representative democracy. These cases explore how corporate responsiveness to community pressures represents an opportunity for Indigenous people to express themselves beyond mainstream political structures, obtaining sought outcomes using extra-parliamentary tactics, such as a sit-in, shareholder activism, media campaigns or tactical use of legislation.

These case studies are only 'flashpoints' in complex ongoing relationships between respective mining companies and local Indigenous communities. They are not presented as definitive accounts, but as illustrations of the potential leverage affected communities might be able to exert over companies and of how impetus for responsiveness gain traction within respective companies. An understanding of why and when companies respond to community demands, appreciating the potential and limitations of community leverage, enables more nuanced analysis of opportunities for social sustainability.

---

[8] Gunningham, Kagan and Thornton (2002) therefore find the concept of a multifaceted licence to operate that incorporates economic, regulatory and social pressures a more useful concept than a single aspect licence.

Century Mine

In the Gulf of Carpentaria in Queensland many Indigenous people have low formal education and suffer poor health, with one of the lowest life expectancies on average in Australia.[9] Until zinc mining began at Century Mine, most employment was through the Community Development Employment Projects (CDEP) scheme (Martin 1998b; Williams 1999). Gulf Indigenous communities have, however, to some degree been able to hold the operators of Century mine to account and attain certain concessions from the miner as a result.

In the 1990s local Indigenous communities, and in particular one charismatic individual accepted by many as representative of broader community views, were able to impel the prospective developer and operator of Century mine, CRA (Rio Tinto's predecessor in Australia), to the negotiating table. Early in negotiations for the mine's development the High Court of Australia, in what became known as the Mabo decision, ruled that Indigenous Australians had common law native title rights over unalienated land over which they could demonstrate continuous connection. Subsequent Commonwealth legislation—the *Native Title Act 1993*—gave registered native title holders a 'right to negotiate', though not a right of veto, over mining (see Altman, Chapter 2; Rowse 2002). For a considerable period after 1993, the native title status over land containing the zinc deposit was unclear, hence CRA lacked a definitive legal license to develop Century, elevating the necessity of community agreement necessary for the mine's operation. Century's management also recognised that following instances of 'radical and aggressive community opposition' to Century's development, even if they secured legal permission to mine, without a social licence to operate the project would remain vulnerable to sabotage.[10]

The Queensland Government strongly supported the project, given Century's employment and regional economic importance. During a difficult point in negotiations, it even offered legislation to by-pass Commonwealth native title legislation, which would enable the mine to proceed, regardless of any claims Indigenous people might have under native title (Cook 1997). This option was refused by CRA which, as outlined below, had come to understand that such litigious tactics did not advance harmonious community relationships.

The emerging corporate strategy to seek community support came, however, from senior employees on the ground, rather than any directive from CRA headquarters. Recognition of Indigenous interests, and the potential implications of Indigenous opposition to mine development, was brought to Century by

---

[9] For example, Mornington Island and Doomadgee have very low environmental health infrastructure, and lack sufficient housing and water supplies (T. Koch, 'Murrandoo, the man', *The Courier Mail*, 23 November 2002, p.35; Martin 1998b).

[10] I. Williams, former Executive General Manager—Mining, Pasminco Ltd, pers. int., 4 August 2003.

Century's Managing Director based on his prior experience with Hamersley Iron's Marandoo Mine in Western Australia (discussed below).

During a protracted negotiation period company negotiators increased an initial offer to local Indigenous communities of $70 000 cash to an eventual $60 million package. The bargaining position utilised by Indigenous interests—by virtue of native title legal provisions, delay tactics and the use of public forums—rendered those seeking to 'regulate' the miner visible to company managers and prominent in company decision-making equations (Trebeck 2005; Trigger 1998). Delays impinging on commercial standing and threats to corporate reputation demonstrated that when Indigenous communities comprehend and penetrate contexts in which large companies operate, they can press for certain corporate behaviour and demonstrable community benefit.

In 1997 representatives from registered native title claimants, the Queensland Government, and Century Zinc Mine signed the Gulf Communities Agreement (GCA). The GCA enabled the mine to go ahead despite outstanding native title claims over the area. It committed the owner of Century to spend $60 million (in 1997 terms) over the life of the mine on employment and training, community development, and payments to native title parties. It pledged native title groups to allow Century's operation in return.

Following the signing of the GCA, CRA (by then Rio Tinto) sold the mine to Pasminco, and Pasminco went into voluntary administration in September 2001 (eventually emerging in 2004 as Zinifex following significant balance sheet adjustments). With the financial problems that preceded administration, Pasminco's concern with community relations at Century lessened – the short term priority of financial sustainability ostensibly assumed priority, to the exclusion of other aspects of sustainability. Insufficient effort and resources were deployed to satisfy the expectations of local Indigenous communities, and Pasminco's focus was limited to compliance with strict contractual aspects of the GCA. Responsibility for GCA implementation was subordinated to a unit within Century's Human Resources. It seems that as Pasminco's financial concerns dominated, community relations suffered. Key individuals supporting community relations effort had left, been retrenched or were largely unable to achieve traction in a company where individuals with authority to affect substantial change no longer appreciated the relevance of the business case for sound community relations.

Nonetheless, prior to November 2002, Century believed it had developed 'good relationships with Indigenous people in the Gulf' (Hall and Driver 2002; Pasminco n.d.; Pasminco Century Mine 2001). In terms of Indigenous employment, Century has been the 'star performer' of the Australian mining industry: over 20 per cent of Century's workforce is Indigenous, the vast majority of whom are local (Pasminco Century Mine 2003).

For many local Indigenous people though, the mine was not seen as delivering benefit to the region beyond employment of individuals (Hall and Driver 2002).[11] There were a number of further complaints, but efforts to convey them to the company proved futile. In November 2002, between 50 and 80 people, mainly residents of Doomadgee (the Indigenous community closest to the mine), occupied the mine site kitchen. The protest did not halt production, but did constitute both an inconvenience and an underlying threat that the action could escalate and impede operations, a particularly ominous consequence given Pasminco's then perilous financial circumstances. After nine days an end to the sit-in was negotiated between those involved, Pasminco and the Queensland Government.

There seems to be recognition from both Indigenous people involved and Century that there has been positive change in the relationship between communities and company since the sit-in.[12] One observer states that the sit-in gave the company a 'wake-up call' that it needed to pay more attention to community issues.[13] It highlighted the repercussions of not being sufficiently proactive in community relationships. Since the sit-in '[Century's] eyes are back on the ball, they now appreciate that the GCA is their "licence to operate", and if they do not sufficiently implement the GCA according to community expectations they will lose this licence'.[14]

Hamersley Iron

In the late 1980s, Hamersley Iron ('Hamersley' now Pilbara Iron), a Rio Tinto subsidiary based in the Pilbara region of Western Australia, sought to develop the Marandoo iron ore deposit located within the Hamersley National Park,[15] claiming that Marandoo was imperative to maintain supply of ore.[16] Despite lacking the trust of many members of Indigenous communities in the Pilbara, Hamersley pursued development of the mine regardless. At this time, prior to the Mabo decision and the *Native Title Act 1993*, Hamersley used adversarial

---

[11] D. Aplin, Doomadgee resident, member of GADC, pers. int., 10 July 2003; A. Chong, former Roche Eltin Joint Venture employee, Burketown resident, pers. int., 10 July 2003; D. Rose, General Manager Pasminco Century Mine, pers. int., 3 July 2003; C. Waldron, Doomadgee resident, pers. int., 10 July 2003.

[12] D. Aplin, Doomadgee resident and member of GADC, pers. int., 10 July 2003; P. Cameron, Gulf region community liaison officer Pasminco Century Mine, pers. int., 13 July 2003; F. Pascoe, Indigenous Business Leader Aboriginal Development Benefits Trust Pty Ltd, Former GCA Manager, Normanton resident, pers. int., 11 July 2003; K. Quigley, Manager External Relations Pasminco Century Mine, pers. int., 1 July 2003; A. Waldron Doomadgee resident, pers. int., 10 July 2003.

[13] C. Reading, Community Development Officer Pasminco Century Mine, pers. int., 2 July 2003; also J. Green, Carpentaria Land Council Doomadgee Office, pers. int., 10 July 2003; D. Rose, General Manager Pasminco Century Mine, pers. int., 3 July 2003; S. 'Bull' Yanner, GCA Superintendent Pasminco Century Mine, pers. int., 2 July 2003.

[14] F. Pascoe, Indigenous Business Leader Aboriginal Development Benefits Trust Pty Ltd, Former GCA Manager and Normanton resident, pers. int., 11 July 2003.

[15] Now the Karijini National Park.

[16] See B. Hextall, 'CRA clear to mine in national park', *The Sydney Morning Herald*, 16 November 1990; M. Stevens, 'Showdown at Marandoo', *Business Review Weekly*, 6 December 1991.

and litigious tactics to progress Marandoo's development, including a media campaign to pressure the Western Australia Government to speed up approval processes.[17] Marandoo was complicated by its location within a National Park and environmentalist's concerns about endangering a rare species of mouse.[18] An alliance of sorts arose between environmental groups and Indigenous interests in opposition to the mine.[19]

Hamersley could not proceed with development until the project obtained clearance under Western Australia's *Aboriginal Heritage Act 1972*. Doing so required identification of Indigenous sites to prevent or minimise any site disturbance. Hamersley asserted that anthropological and archaeological studies conducted since 1974 found no significant Indigenous sites, and when taking ownership of the Marandoo area in the early 1980s it also acquired clearances obtained by the previous owners.[20] Hamersley refused to conduct another survey of Indigenous sites at Marandoo. The Western Australian Government eventually commissioned a further study that found four sites of cultural significance, two of which were located on top of the ore body.

The demand for further surveys and compliance with the *Aboriginal Heritage Act 1972* had significantly stalled Hamersley's intended development time frame. Rather than seeking to prevent the mine altogether, the main objective of many members of local Indigenous communities largely encompassed attaining recognition in negotiations about how Marandoo proceeded and benefits offered to those affected (G. Benn pers. comm. 2003). Delaying Marandoo was a strategy to pressure Hamersley to acknowledge Indigenous interests, enabling the communities to make demands such as increased Indigenous employment in Hamersley's operations. Heritage legislation was the best available (legal) 'device to achieve this'.[21]

Despite resistance and concern in the local Indigenous communities about the destruction of sites, the Western Australian Government passed the *Aboriginal*

---

[17] G. Benn, Aboriginal Legal Service, pers. comm., 5 August 2003; B. Larson, General Manager External Affairs Hamersley Iron, pers. int., 27 March 2003; J. van de Bund, former Manager Aboriginal Training and Liaison, Hamersley Iron, pers. int., 30 June 2003; J. Watson, Western Australian Minister for Aboriginal Affairs 1991–93, pers. int., 5 August 2003; I. Williams, former Group Project Leader Future Mine Resources, Hamersley Iron, pers. int., 4 August 2003.

[18] H. Rumley, anthropologist, pers. int., 27 March 2003; M. Steketee, 'On the fast track to a dispute over Marandoo', *The Sydney Morning Herald*, 5 November 1991; I. Williams, former Group Project Leader Future Mine Resources, Hamersley Iron, pers. int., 4 August 2003.

[19] T. 'Slim' Parker, at Tom Price, pers. int., 3 April 2003; G. Benn, Aboriginal Legal Service, pers. comm., 5 August 2003; J. Pettigrew, Western Australian branch of the Australian Conservation Foundation, quoted in M. Steketee, 'On the fast track to a dispute over Marandoo', *The Sydney Morning Herald*, 5 November 1991; C. Smith, Community Development Hamersley Iron, pers. int., 2 April 2003; J. Watson, Western Australian Minister for Aboriginal Affairs 1991–93, pers. int., 5 August 2003.

[20] T. Darvall, 'Hamersley in Ad Offensive for Iron Project', *The Age*, 2 August 1991.

[21] T. 'Slim' Parker, pers. int., 3 April 2003; J. Watson, Western Australian Minister for Aboriginal Affairs 1991–93, pers. int., 5 August 2003; I. Williams, former Group Project Leader Future Mine Resources, Hamersley Iron, pers. int., 4 August 2003.

Heritage (Marandoo) Act 1992 that excluded the area from the *Aboriginal Heritage Act 1972*. The mine was constructed after a delay of two years and substantial legal cost to both the company and local communities (Bradshaw 1998; Eggleston 2002). Pertinently, Hamersley staff have since described the protracted processes of pursuing development as harrowing, expensive and significant enough to change the company's approach to community relations.[22] Key company individuals realised such disputes are not good for long-term business activities and future mining activity in the Pilbara.

Instead, seeking to improve community relationships and prevent any future delay to its development plans, Hamersley established an Aboriginal Training and Liaison Unit in 1992 to increase Indigenous employment and repair company relationships with Indigenous people in the Pilbara. When it came to the Yandicoogina mine in 1997, in the post-native title climate, Hamersley undertook to negotiate and reach agreement with Indigenous representatives, rather than the legalistic and confrontational approach that characterised Hamersley's effort to develop Marandoo a few years earlier.

Hamersley now states that it recognises the existence of native title and the interest it bestows on Indigenous people as 'real stakeholders' in lands Hamersley uses for mining (van de Bund 1996). In recognising certain demands and expectations of local Indigenous communities, Hamersley aims 'to meet its strategic goals of maintaining access to land and resources and having the necessary flexibility in [its] operating environment' (Hamersley Iron 2000). It has been Indigenous leverage, potently demonstrated in the delay of Marandoo, which drove this strategy of responsiveness, reinforced by changes to community relations strategies adopted by Hamersley's parent company, Rio Tinto.

Rio Tinto's Indigenous relations

Changes in Hamersley's approach to external stakeholders were given added momentum by support from senior managers within Hamersley's parent company, Rio Tinto. Key managers discerned the potential benefits from improved Indigenous relations, and were able to shift company policy towards one more responsive to the expectations of relevant Indigenous entities. Rio Tinto's transformation from campaigning to minimise recognition of Indigenous rights (see for example Howitt 1998), to seeking a reputation as Indigenous Australia's preferred development partner, has been self-described as a 'philosophical sea-change' (Cusack 2001). The Marandoo dispute highlighted the prudence of this new strategy, augmented by the advent of native title. Agreements with Indigenous interests being a matter of course in Rio Tinto

---

[22] D. Trigger, Anthropology Department, University of Western Australia, pers. int., 25 March 2003; B. Larson, General Manager External Affairs Hamersley Iron, pers. int., 27 March 2003; J. van de Bund, former Manager Aboriginal Training and Liaison, Hamersley Iron, pers. int., 30 June 2003; see also Rio Tinto Limited 2001.

operations overseas, rising risks to reputation through non-governmental organisation scrutiny, and violent physical protest and subsequent premature closure of Rio Tinto's (then CRA) Bougainville Copper Mine in Papua New Guinea cumulatively demonstrated the prudence of increased community relations effort (see Trebeck 2005). As Rio Tinto's Chief Financial Officer explained, 'the business case [for corporate social responsibility] rests first on risk mitigation. Without corporate social responsibility, a mining company may eventually be unwelcome as an investor in its host environment'.[23]

Individual employees appreciated the imperative for change and drove its initiation. For example, the leadership of Rio Tinto's then Chief Executive Officer, Leon Davis, is considered crucial in initiating change and enabling implementation of new community relations policies. Davis has been described as 'spearheading' the company's decision to shift from legal wrangling to negotiations with Indigenous people.[24] His seniority in the company hierarchy presented an ability to effect change quickly, underpinned by an appreciation of the business case and personal commitment to harmonious relations with Indigenous Australians.

In particular, Davis has been described as 'that way inclined anyway' towards supporting positive outcomes for Indigenous people.[25] His efforts in Indigenous relations were in part premised on his long-held beliefs that companies can deliver social benefit, and that Indigenous Australians deserved better treatment from mining companies. His career included work at sites overseas where engagement and deals with local people were routine, and also at Bougainville in Papua New Guinea where hostility of local people forced early mine closure in 1989. Davis describes

> the loss of the [Bougainville] mine as a huge shock. It was an awful thing to happen. BCL was a huge investment, and the diamond in the CRA crown. Bougainville was a lesson I learnt and which I took with me.[26]

Following advocacy of the prudence of corporate social responsibility by Davis and others with similar perceptions, Rio Tinto's new position regarding Indigenous relations was publicly stated in 1995. In a landmark speech, Davis articulated 'satisfaction' with the main tenets of native title legislation, and outlined Rio Tinto's changed approach towards Indigenous communities near its operations: away from a litigious stance towards development of mechanisms to share with and compensate Indigenous people for mining activity on their land (Davis 1995). Davis describes this move as 'hard headed'.[27]

---

[23] G. Elliot, Chief Financial Officer, Rio Tinto, quoted in World Economic Forum 2004.

[24] P. Manning, 'Poor fellow mining country', *Sydney Morning Herald*, 18 April 2003.

[25] I. Williams, former Executive General Manager—Mining, Pasminco Ltd, pers. int., 4 August 2003.

[26] L. Davis, pers. int., 14 February 2005.

[27] L. Davis, pers. int., 14 February 2005; Davis 2002.

Rio Tinto business units were told by the company's Head Office that they needed to alter their Indigenous relations approach and practice. New strategies were mandatory, but should be structured according to local circumstances. Middle managers, facing pressures of output quotas, make decisions according to company policy and incentive structures, underlining the importance of high-level directives in realising internal change. Company policies and mandates bridge any perceived short-term costs (immediate outlays for developing community relations), with the long-term corporate gains (enhanced company reputation and improved land access). Public declaration of the need to achieve positive relationships with Indigenous people also aided Rio Tinto employees already predisposed to more conciliatory external relations.

The story, of course, is not one of universal acceptance and smooth implementation. The external face of Rio Tinto's corporate social responsibility in the form of publicised programs and publications masked internal dissent and scepticism. For example, when Rio Tinto's Indigenous employment program was introduced at company mine sites, there was little enthusiasm for the new strategy and the effort it required (Mays 2003). Despite high-level decrees from those charged with management of Rio Tinto's external operating context, many mid-level employees retained the perception that technical competence is the main ingredient for corporate and personal success. Perception of the long-term corporate interests (which often justifies community engagement) seems fundamental to appreciating success factors as existing beyond short-term budgetary objectives.

Jabiluka

Rio Tinto's changing approach towards responsiveness to those with capacity to impact mine operations constitutes the background in which Rio Tinto acquired North Limited in 2000. North owned 68.4 per cent of Energy Resources of Australia (ERA), which owned the Jabiluka uranium mineral lease in the Northern Territory. The Jabiluka example illustrates the sorts of circumstances that might prevent mining altogether. In seeking to attain a social licence to operate, the language of 'win-win' outcomes is often used by miners to justify their presence and promote what they bring to a community. This assumes that mining will proceed and that local communities can derive outcomes sufficiently beneficial they can be described as a 'win' for the communities. Such language and its implication leaves little room for communities to reject mining outright. The campaign against development of Jabiluka illustrates how such a rejection can be enforced—that is, how the social licence to operate can be withheld (see also Trebeck 2007a, 2009).

The anti-Jabiluka campaign involved the confluence of three issues: local Indigenous opposition, Jabiluka's location surrounded by the World Heritage listed Kakadu National Park, and concerns over uranium mining. Protest by campaigners for Indigenous land rights, environmentalists, and anti-nuclear activists encompassed legal action, efforts at education, mobilisation of national and international opposition, physical demonstrations, shareholder activism and parliamentary lobbying. The culmination of these components of the campaign against Jabiluka meant that a seemingly hopeless situation faced by a small and historically marginalised community was transcended. Jabiluka became an iconic struggle for anti-uranium stalwarts, but anti-uranium and environmental opposition to Jabiluka derived much potency from the discontent of traditional owners—the Mirrar people. Key traditional owners of the Jabiluka prospect have been unambiguous in their opposition to Jabiluka's development in recent times, opposition informed by concerns over the disturbance of sacred sites and experience of the nearby Ranger uranium mine where social and environmental impacts are considered adverse.

Campaigning by the Senior Traditional Owner, Yvonne Margarula, and the then Executive Officer of the Gundjehmi Aboriginal Corporation, Jacqui Katona, mobilised the international non-governmental organisation community, and led institutions such as the European Parliament and the United States Congress to pass resolutions against Jabiluka's development. In 1998 Margarula attended the United Nations Educational, Scientific and Cultural Organisation (UNESCO) World Heritage Bureau meeting in Paris, instigating a UNESCO Mission to Kakadu. The mission reported to the United Nations World Heritage Committee that there were 'significant ascertained and potential threats' to the National Park associated with development of Jabiluka, and recommended construction work at Jabiluka be halted.

The Australian Government refused to adopt these recommendations and the Department of Foreign Affairs reportedly spent six months and over $1 million lobbying the United Nations World Heritage Committee and key decision-makers in its member countries, ultimately leading to the United Nations not placing Kakadu National Park on the endangered list (Banerjee 2000).

Jabiluka was opposed by many of Australia's environmental organisations, church, trade union and community groups. The Australian Senate passed a resolution opposing Jabiluka's development. An opinion poll in 1998 found that 67 per cent of Australians were against mining at Jabiluka.[28] In urban centres protest against Jabiluka grew, including large rallies and public events in all capital and many regional cities. In March 1998 a blockade was established near the Jabiluka site, lasting eight months and involving over 5000 people. More than 500 protestors were arrested for trespass, including traditional owner Yvonne Margarula. The blockade meant that the company could not access the mine site as desired, despite being given legislative permission to mine and

---

[28] Quoted in C. Miller, 'North shareholders protest over Jabiluka', *The Age*, 4 June 1999, p. 7.

supported by strong government backing. By flying employees and equipment to the site by helicopter ERA was, however, able to complete construction of the mine decline and retention pond in 1999.

ERA and its parent company North thus faced prolonged opposition to development of the Jabiluka mine. Annual General Meetings and company headquarters became 'combat zones' where anti-Jabiluka campaigners targeted shareholders and company management.[29] In March 1999, the Melbourne premises of North were blockaded for four days. Activists also lobbied major North investors directly. Various financial institutions were sent anti-Jabiluka material, and in 1998 there was a national day of action targeting North's main bank, Westpac, for its involvement. By June 1999 almost $7 million worth of North shares had been divested. Insurance company NRMA, for example, stated publicly that the reason for selling its North shares was concern about Jabiluka, deeming the project to be financially, politically and environmentally risky.[30]

Alongside the disinvestment by some financial institutions, there was strident shareholder activism. The Wilderness Society obtained North's share register and contacted over 67 000 shareholders, informing them about Jabiluka and instigating the formation of a shareholder pressure group, North Ethical Shareholders. In October 1999 an Extraordinary General Meeting was held at the instigation of North Ethical Shareholders. Although the proposals put by the North Ethical Shareholders were defeated, the event attracted significant negative publicity for North and required diversion of management time to dealing with the activists.[31]

It seems, given North's strident pursuit of Jabiluka's development and its reaction to protests, that the company was almost impervious to many elements of the anti-Jabiluka campaign. Construction at the Jabiluka mine took place regardless of traditional owner objections. Arguably, if elements of the campaign such as urban protest and international institutions opposing the mine did not achieve their objective, they at least bolstered traditional owners in their manoeuvring against Jabiluka.

A significant lever that traditional owners wielded was the location of the mill for Jabiluka ore and an access road between nearby Ranger and Jabiluka. Traditional owners, by virtue of a clause in a 1991 lease transfer agreement (transferring ownership from Pancontinental to ERA), held an effective veto over milling of Jabiluka ore at the Ranger mill. The cost of building a mill to process Jabiluka ore at the Jabiluka site was over $200 million, a cost ERA hoped to avoid by trucking ore to the existing Ranger mill 22 kilometres away. In the

---

[29] J. Rose, pers. int., 29 March 2004; D. Sweeney, anti-nuclear campaigner, Australian Conservation Foundation, pers. int., 1 April 2004.
[30] Cited in The Wilderness Society and The Mineral Policy Institute 1998.
[31] J. Rose, 'Blame it on the Rio Tinto moment', *The Australian Financial Review*, 12–16 April 2001, p.10.

context of then poor uranium prices, the veto over doing so, effectively necessitating construction of a new mill at Jabiluka, made the economics of the Jabiluka project far less attractive to ERA and North.

There were two phases in the campaign to stop development of Jabiluka—before and after 2000 when Rio Tinto acquired North, and with it majority ownership of ERA. Rio Tinto would seek to avoid mining without community consent, whereas North seemed to rely on government sanction to mine, regardless of whether local communities sanctioned this or not. Rio Tinto's position was that although the Northern Territory and Commonwealth Governments supported the mine, this was insufficient, as consent was also needed from traditional owners. Paradoxically perhaps, being a large international company with an important brand, making a concerted public effort to move away from historic poor performance, and seeking to cultivate positive reputation amongst Indigenous communities more broadly, Rio Tinto was seemingly unwilling to weather the reputational attacks that North ostensibly tolerated. The potential cost to Rio Tinto's reputation of proceeding with Jabiluka in the face of local opposition was increasingly evident as adverse market conditions for uranium in the few years after acquiring ERA made Jabiluka relatively marginal to Rio Tinto's overall commercial strategy—as one observer said, the Rio Tinto brand was 'suffering much grief for a little mine'.[32] For example, in September 2002 Rio Tinto's stake in ERA was worth $172 million, less than 1 per cent of Rio Tinto's total net present value.[33]

In late March 2001, Rio Tinto's Chief Executive Officer indicated that the company would not support development of Jabiluka unless there was a substantial alteration of community attitudes, alongside improved uranium prices (Clifford 2001).[34] This moderate concession did not placate the anti-Jabiluka campaign. At Rio Tinto's Australian Annual General Meeting the following month protestors held a 'die-in' and company shareholders were forced to step over 'dying' protestors.[35] At Rio Tinto's London Annual General Meeting the next year, a statement on behalf of traditional owners was presented, rejecting mining on Mirrar land and demanding that Rio Tinto leave the area. In September 2002 Rio Tinto Chairman Sir Robert Wilson strengthened his company's position not to mine Jabiluka without acceptance from traditional owners; he also stated, for the first time, that the site would be rehabilitated and the mine's entrance

---

[32] D. Sweeney, anti-nuclear campaigner, Australian Conservation Foundation, pers. int., 1 April 2004.

[33] B. Hextall, 'Rio keeps door open', *The Australian Financial Review*, 6 September 2002, p. 65.

[34] Wilson, for example, highlighted the lack of viability of Jabiluka mine as an important consideration—admitting 'we don't see the development as being viable in any case' (quoted in I. Howarth, 'Rio Tinto concedes defeat on Jabiluka', *The Australian Financial Review* 4 April 2002, p. 59; and J. Rose, 'Blame it on the Rio Tinto moment', *The Australian Financial Review*, 12–16 April 2001, p.10).

[35] Organisations included Friends of the Earth, the Mineral Policy Institute, Greenpeace, and the Australian Conservation Foundation.

sealed.[36] Until this declaration, the Mirrar had understood that despite their protests the Jabiluka mine could still go ahead without their approval once Ranger was no longer operational. An agreement (the Jabiluka Long Term Care and Maintenance Agreement) has subsequently been developed between ERA and traditional owners and their representatives, regarding rehabilitation and maintenance of the Jabiluka site and future consultation over development.

An important element of the Jabiluka campaign, with parallels in both Hamersley Iron's experience at Marandoo and negotiations for Century mine, was the role of government. In each case both state and Commonwealth governments sought to facilitate mine development, including passing special legislation or ignoring United Nations recommendations. Government approved development of Jabiluka in spite of protests by traditional owners, as well as wider opposition. Government support for Jabiluka has been on 'national interest' grounds, with the Prime Minister citing the imperatives of globalisation as necessitating Jabiluka's development.[37]

Some commentators have portrayed Rio Tinto and ERA's position of not mining without local community consent, and the anti-Jabiluka campaign that drove this strategy, as a 'David and Goliath' win for the Mirrar people. Katona observes that it demonstrates how:

> a bunch of women from the bush, the majority of whom were illiterate, spoke English as a second or third language, stuck to their guns, and took on the mining company, and hunted them down … Wherever the mining company was speaking publicly we were able to have agents or representations made … directly to the face of the company to … let them know that we were not going to be beaten....[38]

It shows what leverage is necessary to capture corporate attention and response: ability to impact the commercial position of the company, whether in the short term through economic levers or over the long term via influence on company reputation.

Discussion and conclusion

These case studies suggest that uptake of corporate social responsibility depends on responsive individuals in the company, especially senior management, being receptive to the pressures for responsiveness; namely those that appeal to commercial interest. Once a company has accorded a community group or organisation 'stakeholder' status, the degree to which their demands are acknowledged and addressed is a function of how managers perceive stakeholder

---

[36] 'Jabiluka in Mirrar hands', *The Northern Star*, 6 September 2002, p. 23.
[37] Prime Minister Howard cited in G. Milne, 'Howard's in the red over Brown and Greens', *The Australian*, 7 December 1998, p. 15.
[38] ABC Radio National, *Late Night Live,* 21 August 2003.

power (to impact on the firm), the legitimacy (of the group or its claim) and urgency with which the demands are pressed (Agle, Mitchell and Sonnenfeld 1999; Mitchell, Agle and Wood 1997). While urgency seems difficult to delineate from power (beyond how acutely demands for response impact on shareholder value), responses seen in the case studies suggests that power is the most relevant element in this equation. It was those communities able to affect the company's capacity to operate which were accorded most corporate attention. The immediate response to the sit-in at Century mine, after little response from the company to written attempts at communicating aspirations, starkly demonstrates this, as do the various elements of the anti-Jabiluka campaign.

Moreover, comparison of North and Rio Tinto's handling of the anti-Jabiluka campaign, and in particular reaction to traditional owner opposition, illustrates divergent criteria for stakeholder legitimacy: North adopted Western-centric 'majoritarian' notions of who speaks for land (Trebeck 2005), while Rio Tinto accorded value to authority based on traditional ownership according to Indigenous political systems and sought to respond to the concerns of these community groups in particular.

The case studies also demonstrated how interpretation of community demands for corporate social responsibility is often shaped by the values held by a corporate executive, or their experience of a crisis where the company's social licence to operate was jeopardised (Agle, Mitchell and Sonnenfeld 1999; Parker 2002; Webley 2001). Orlitzky and Swanson (2002) model this as how 'attuned' executives are to the demands of relevant stakeholders. For example, when a crisis (such as delay of Marandoo or closure of Bougainville mine) demonstrates capacity of certain stakeholders to impede company operation, individuals within the company who are pre-disposed to corporate social responsibility, even for 'moral' reasons, gain leverage for their views, while recalcitrants might come to appreciate the business case for it. Personal commitment to corporate social responsibility for moral reasons thus co-exists alongside the business case that drives prudent support for corporate social responsibility.

While a moral motivation might be sufficient in some cases and for some individuals, because the raison d'être of companies is profit and commercial continuity there is a need to understand and communicate the business case for corporate social responsibility internally to achieve change in corporate responsiveness to community demands. Only when the strength of a business case is evident will resources be dedicated to implementing corporate social responsibility (McLaren 2002; Parker 2002). Thus, if an individual or group is completely ineffectual in relation to a company, then there is unlikely to be a persuasive business case for the company to respond to the interests of these entities, save for a sense of moral duty held by some individuals within a firm.

The examples of Century, Hamersley Iron, Rio Tinto and the anti-Jabiluka campaign support the contention that key individuals within organisations, notwithstanding any personal moral motivations, ultimately rely on business case arguments to gain internal traction for change and operational support for corporate social responsibility initiatives. In turn, however, the effectiveness of otherwise austere structures and policies comes from individuals. These individuals may be driven by either an appreciation of the business case, company policies, promotion and pay incentives, or moral commitment to improved Indigenous relations (Trebeck 2005).

That civil regulation often hinges on key individuals can, potentially, constitute risk. When such (arguably 'enlightened') individuals leave or when incidents that initially highlighted the need for corporate social responsibility recede in relative prominence in management attention then the drive and vigour of implementation is likely to abate. For example, for a time corporate social responsibility was ostensibly deemed relatively less important than the eventual catastrophic financial situation of Century's parent company. Developing structures to internalise corporate social responsibility might reduce the burden of advocacy away from a few individuals. Equally, however, structures themselves are insufficient: frameworks and programs to deliver corporate social responsibility are of little use if those charged with implementation do not do so with enthusiasm, understanding and appreciation of the necessity of their task. This was evident in the way some committees created to implement the GCA at Century were mismanaged, becoming counter-productive and a cause of community discontent that eventually impelled, in part, the 2002 sit-in (Trebeck 2005).

Civil regulation then is successful when community actions (or deliberate inactions) recognise that the most effective way in which to shape company behaviour is via financial gain or loss, thereby creating a business case for desired corporate change. It is individuals in the company who must recognise this business case, as occurred when management attention to the need for sound community relations became focused by incidents such as Bougainville's forced closure, the Marandoo dispute, the advent of native title, Century's sit-in and the anti-Jabiluka campaign. Here, a shift in the balance of external influences brings Indigenous demands to the fore by—directly or indirectly—threatening financial performance. When such civil regulation consequently gains corporate response there has invariably been a convergence of personal commitment and commercial imperative.

Given that companies are neither uncomplicated nor internally homogeneous, community pressures for corporate social responsibility will affect respective elements of a company differently. Adding to this complexity, both the organisation itself and individuals within it, function within a diverse social system (Keskinen, Aaltonen and Mitleton-Kelly 2003). Companies are porous entities, with individual employees having their own external networks that inform their actions and motivations. Once the objectives of individuals within companies are accumulated, and external stakeholders accounted for, companies can be seen as assemblies of relationships with respective audiences, requiring specific corporate responses and actions. This was evident, for example, in the divergent expectations of corporate behaviour held by the financial sector, compared with those of local communities, seen during Pasminco's financial difficulties.

There are frequently structural causes of such divergence. Employees at mine sites—where much corporate social responsibility is made manifest—often have strict job requirements that encompass output quotas, directing their priorities towards immediate production targets. Business units are faced with contradictory signals from headquarters, with consequent dilemmas regarding priorities: they must deliver production targets, reduction of costs and manage industrial relations, while also being expected to implement more intangible, costly community relations initiatives with their inherently longer-term outcomes. Time and production constraints in pursuit of profit often override other pressures. If corporate social responsibility initiatives are deemed expensive relative to other objectives, they are unlikely to be entered into, unless mandated by headquarters. Those at headquarters are charged with navigating the company's external environment for the longevity of shareholder value, whereas managers at mine sites appreciate less the geographical cross-subsidisation of reputational capital.

In addition, an essential element in the notion of civil regulation is the section of civil society doing the 'regulating'. Findings from the case studies suggest that there are several characteristics of civil society that might undermine the capacity of civil regulation to deliver beyond isolated instances. The representativeness of civil society—how those organisations demanding change from companies reflect the actual needs and demands of those in whose interests they purport to act—is vital. The greater the deficit in organisations' representativeness, the greater the likelihood that results of civil regulation will be skewed away from the interests of those affected. Such distortion, where outcomes do not reflect the actual wishes and expectations of communities, undermines any advancement of citizen sovereignty over companies.

The nature of the 'civil regulators' is an area for further research. It is worth suggesting, however, that it is in the company's interest to understand how any organisation actually relates to and represents its constituents—relationships will ultimately be undermined if citizens feel they are disenfranchised by engagement taking place between companies and certain elements of civil society. The structure of Century's GCA reflects dilemmas associated with civil society organisations. Part of the discontent that led to the 2002 sit-in at Century Mine was a feeling amongst some local Indigenous community members that they had not been sufficiently accounted for in the GCA, referring to themselves as 'the forgotten Waanyi' and eventually seeking alternative means (the sit-in) by which to achieve their demands.

Given these (and other) caveats, civil regulation and increased community participation in corporate decision-making is insufficient to attain the stated aims of particular communities (Bendell 2000a; Trebeck 2005; see also Maddox 1991). This limits the scope for social sustainability potentially realised through corporate social responsibility. Moreover, necessity of the business case highlights a role for civil society in maintaining vigilance and sustaining the context that prompts companies to consider communities. Inevitably the capacity of civil society to sustain this vigilance will be shaped by factors underpinning a community's sustainability, while corporate social responsibility itself can increase the sustainability of a community (see Trebeck 2007b). Emerging from this complex multi-directional relationship is evidence that the success of civil regulation to deliver community wishes in some contexts, as illustrated above, does illustrate that social sustainability can be advanced by utilising various levers to alter corporate operating frameworks.

7. Indigenous entrepreneurialism and mining land use agreements

Sarah Holcombe

Introduction

In the Pilbara region of Western Australia, the focus of this chapter, the mining boom—or the 'ramp up' in production as it is referred to within the industry—is such that negotiations for land access have intensified and annual payments to the Indigenous organisations examined here have increased threefold since 1997. These organisations are the Gumala Aboriginal Corporation (Gumala) and Gumala Investments Pty Ltd (GIPL); set up to manage the Yandicoogina Agreement (YLUA). This chapter critically examines the mechanisms through which Indigenous beneficiaries are able to articulate to a Land Use Agreement (LUA) as individuals, with specific attention to possibilities for entrepreneurial activity. The range of possibilities for direct and tangible benefits from agreements is a key area of Indigenous concern. Because of the complexity of the land use agreements examined for this project, this chapter will only focus on Rio Tinto's YLUA in the Hamersley Ranges of the Pilbara. Indigenous 'beneficiaries' operate in a politically volatile and economically expansive context in this area of engagement with the mining sector and the organisations developed to manage the agreements.[1]

To closely consider a range of articulations with the YLUA, it seems apposite to provide a series of examples of the ways in which particular individuals have engaged with opportunities and, conversely, individual critiques of structural limitations on engagement within the YLUA. To some extent, these examples will be biographical. Indeed, a few Indigenous people in the Hamersley region have published biographies, several with the assistance of Rio Tinto and organisations established to manage agreements. One of the biographical 'case studies' discussed in this chapter will be that of Lola Young (2007). The other individuals discussed will, for ethical reasons and to retain anonymity, be referred

[1] I would like to thank members of Gumala based in Tom Price and the surrounding homelands, and those members residing in Port Hedland, Wickham, Roeburne and Karratha who spent time speaking with me for this project during 2003 and 2004; with special thanks to Darren Inji, and Slim Parker. Thanks also to staff in the Tom Price Gumala office, especially Bill Day (during that time and more recently) and Larry Softley, who supported this project during field research. Thanks also to Mark Simpson and staff of Rio Tinto's Aboriginal Training and Liaison Unit in Dampier. More recently, discussions have been held with Gumala project officer Don Gordon and the new Gumala Chief Executive Officer Steve Mav, who have been keen to articulate the new generation Gumala that this paper has only begun to grapple with. The views expressed in this paper are not necessarily those of Gumala staff. Finally, I contacted all four of the case study individuals in this chapter, providing drafts to them and best efforts have been made to encourage all of them to vet their stories.

to by pseudonyms. These brief accounts of individual engagements with the industry are included towards the end of the chapter, after consideration of the context within which individuals are enabled to operate. While entrepreneurialism is a focus of this research, it encompasses the ways in which individual beneficiaries to the Agreement can benefit, rather than solely examining possibilities for developing business enterprise.

Fig 7.1 The Pilbara region, Western Australia

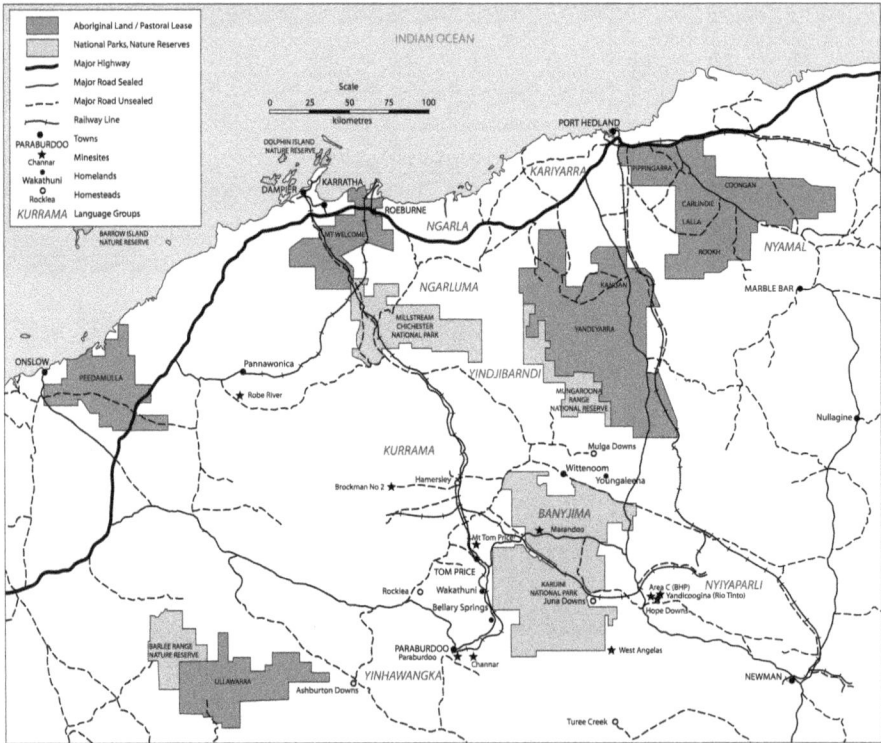

Source: Cartographic Services, Research School of Pacific and Asian Studies, ANU.

There are a range of structures and mechanisms through which Indigenous 'beneficiaries' to the YLUA can engage with its financial benefits. Many of these benefits are indirect or process oriented—ranging from gaining sitting fees from committee membership and, presumably, having some element of investment over allocation of resources, to working on heritage clearances for the Company. While this analysis is concerned with the YLUA, and relevant elements of the Agreement structures will be outlined below, the broader economy of the region is such that individuals may be party to more than one agreement with different mining companies. Indeed, some individuals sit on such a range of committees that their attendance at meetings is a full time occupation. Other more direct and secure means for financially engaging with the YLUA include employment for organisations that were established to manage it; elderly and infirm people gaining regular 'top-up' monies from the Foundation (no longer operating), to being employed on the mine site (though this is open to anyone Indigenous, non-Indigenous, local or global) and in associated operations; working on specific cross-cultural programs provided to non-Indigenous mine staff; and self-employment in small mine-related contracting businesses.

An Indigenous entrepreneurialism

The use of the term 'entrepreneurialism' is informed by Indigenous Canadian experience where social sustainability and the recognition of political rights has been a more evident outcome of entrepreneurial activity than economic sustainability (Peredo et al. 2004). This alternative understanding of entrepreneurialism is useful in analysing the motivations of many Indigenous Australians' attempts at entrepreneurial activity through the land use agreements under consideration here. Non-market driven modes of entrepreneurialism also need to be canvassed. Recent research in Australia by Hindle and Lansdowne (2005), that attempts to draw together Australian and American Indian perspectives on Australian Indigenous entrepreneurialism, also suggests that entrepreneurship here should be understood as encompassing more than economic imperatives. They define Australian Indigenous entrepreneurship as:

the creation, management and development of new ventures by Aboriginal people for the benefit of Aboriginal people. The desired and achieved benefits of venturing can range from the narrow view of economic profit for a single individual to the broad view of multiple, social and economic advantages to entire communities... (Hindle and Lansdowne 2005: 132).[2]

In light of this it is appropriate to reconsider notions of success or failure and indeed the criteria under which 'business' applications are considered—for instance, within the framework of agreements. Certainly this lack of support for developing individual and family businesses was raised as a significant issue for Indigenous beneficiaries of the YLUA in the Review of the associated Gumala Investments Pty Ltd and the Trusts (Hoffmeister 2002). Foley, another researcher in this area in Australia, also found that the definition of success for Indigenous Australian entrepreneurs is 'based on subjective notions derived from the entrepreneurs themselves, and is not solely restricted to financial criteria' (Foley 2006: 1, see also 2003).

---

[2] I am cautious in using this research, however, as it seems to have focused on developing an erratic and confusing method utilising a number of 'theories' whereby a 'formal paradigm' is sought to determine whether entrepreneurial business ventures can claim to be Aboriginal or not. Amongst other issues the claim for 'global relevance' is peculiar as the interviews were restricted to Australia and the United States. No examples of Aboriginal enterprise are canvassed and nor it seemed were Aboriginal enterprise operators spoken with.

Foley focuses on Indigenous entrepreneurs who are the owners and managers of stand-alone commercial enterprises in urban environments because they comprise the majority of Indigenous entrepreneurs in Australia. Although Foley focuses on mainstream commercial success, importantly he problematises the tension within Indigenous policy between community development and the development of the individual (2006: 5). He notes that there is little recognition of the socioeconomic category of the individual Indigenous entrepreneur and that policy funding structures have a blanket approach toward Indigenous 'communities' (Foley 2006: 5–6).[3] Indeed, this issue of the way in which the Indigenous 'community'—as an encompassing category within the YLUA—is embedded in the 'community development' discourse was discussed at length in a previous paper (Holcombe 2006). It is important, nonetheless, to observe this issue again as a restrictive and limiting factor that discourages entrepreneurialism.

There are legal limitations on the Trusts set up under the YLUA and the 'community benefit package' (as the agreement is also referred to) as they impose both corporate expectation and constraint on Indigenous action. This chapter proposes that major resource companies incorporate strategies in land use agreements that enable individual Indigenous people, who are party to agreements, to develop the capability to deal with the potential opportunities that these agreements present. At the same time it proposes a broader definition of 'opportunity'. A central contention is that mobilising individual agency to strengthen capacity to be socially productive may not necessarily lead to mainstream economic productivity, but rather to producing community and family economies. Such local informal economies tend to be marginalised by the dominance of large scale resource development and the focus on mainstream employment in the industry and in government policy (see Holcombe 2006).

How is this active engagement facilitated or hampered by the structures set up to disburse the benefits? Such structures are not neutral frameworks that allow an innocent expression of Indigenous interaction. They are deeply informed by a development discourse and are themselves a major arena of development in remote areas; these structures need to be understood in those terms. According to the United Nations, the fundamental purpose of development is '...to enlarge people's choices. In principle, these choices can be infinite and can change over time... People often value achievements that do not show up at all, or not immediately, in income or growth figures...' (Haq n.d.).

---

[3] Foley focuses on an urban/remote divide: typifying and contrasting urban individuals as against remote communities. However, he tends to use the concept 'community' un-problematically as assumed places of shared interests. I would argue that 'communities' can't be assumed in remote areas either. There is a considerable literature critiquing the community concept, see especially Gusfield (1975) and Cohen (1985) and for (Central) Australia, Holcombe (2004b).

It seems important to recall the fact that, in the Pilbara for instance, only 30 per cent of Indigenous adults participate in the mainstream labor market and, of these, 22 per cent work in the mining industry (Taylor and Scambary 2005: 28–37). These findings suggest that focusing on alternatives may be a pragmatic way to think more broadly about the scope of the YLUA and what it can realistically deliver. Likewise, the encouragement of diverse alternative economies would also seem to be critical to any post-mine life in these remote regions and encouraging entrepreneurs is an element of planning for this future. As human geographers such as Gibson-Graham (2002) and Howitt (2001) have found, alternative economies encompass heterogeneity and run counter to the notion that there is a homogenous community of interests that is served by a uniform capitalist economy (see also Community Economies Collective 2001). Certainly, the manner in which the term 'community' is embedded in the YLUA presumes a unified and shared set of communal Indigenous interests—and consequently the diversity of these interests becomes submerged (see Altman, Chapter 2; Scambary, Chapter 8).

This research is informed by the human development approach pioneered by economist Amartya Sen (Sen and Anand 1994; Sen 1999) and further refined by normative philosopher Martha Nussbaum (2001). This approach has been used as a conceptual framework by the United Nations Development Program since 1990 'to inform policy choices in many areas, from poverty reduction to sustainable development to gender to globalisation to governance' (Fukuda-Parr, Lopes and Malik 2002: 1). Research by the World Institute for Development Economics, especially by Sen (Sen and Anand 1994; Sen 1999) and Nussbaum (2001), has found that the principle of 'each person as an end' needs to be the fundamental basis for 'development planning'. In this human rights based approach to agreements as development, instead of asking questions about people's satisfactions (which are subjective and may be conditioned), the questions are about what they are actually able to do or to be, as it is in this space that social equality and inequality are best raised (Nussbaum 2001: 12). For instance, the dominant approach to assessing quality of life previously focused on Gross National Product (GNP) per capita, 'treating the maximisation of this figure as the most appropriate social goal and basis for cross-cultural comparison'. However, such assessment has to investigate the distribution of this wealth and income; as encapsulated in 'who has got the money and is any of it mine?' (Nussbaum 2001: 60).

This approach to issues of the distribution of wealth and benefits invites examination of the way the $60 million plus benefit package of the YLUA affects individuals: where and how does this money get distributed and can 'beneficiaries' say any of it is theirs or that they had a role in deciding how it was spent or invested? The question that this then begs is whether the organisations set up under the YLUA are creating an enabling environment for the Indigenous stakeholders they are purported to benefit; are they encouraged to be agents?

Weighing up immediate tangible benefit—requests of cash or seed money for a family or individual enterprise—against long term or nascent benefit is a complex balancing act and there are not any simple solutions, or none that are directly transferable across different regions. Nevertheless, this is a tension that has been found to exist in relation to other agreements and patterns have formed around this difficult balance (see Altman, Chapter 2). In the YLUA the balance is heavily weighted in favor of long term future investment and financial risk minimisation, and does not give enough credit to potential or immediate entrepreneurial possibilities or investment in promoting these opportunities. Indeed, this was the major thrust of the findings of the Review of the Trusts and GIPL (Hoffmeister 2002). As will be discussed, however, in the six years since this Review there has been a shift toward diversification and small scale business development. As Gumala—one of the two key organisations set up to manage the agreement—has noted, they are an organisation in the process of transforming themselves.

According to the YLUA not less than 40 per cent of the available income should be invested and not less than 30 per cent spent on education and training, and business development. In fact it was found that most of the available income was invested, approximately 40 per cent more than targeted (Hoffmeister 2002):[4] as of June 2007 there was approximately $40 million being held by GIPL (Bill Day, pers. comm. June 2007). This business philosophy weighted toward long term investment may be, in part, a risk aversion strategy.[5]

However, this approach to investment can limit the potential of the money to build human capital, or, as the first non-Indigenous Director of Gumala Enterprises (GEPL) (and Graham (Polly) Farmer Foundation Chief Executive Officer (CEO)) noted, 'if the money is not being spent it's not doing good' (pers. comm. 2004).[6] Aristotle's prescient observation seems pertinent here, 'wealth is evidently not the good we are seeking; for it is merely useful and for the sake of something else' (Aristotle 350 BC). The 'something else' may be understood through the concept of fungibility, recently bought to prominence in Australia by Noel Pearson (Pearson and Kostadikas-Lianos 2004; see also Bourdieu 1986; Coleman 1988)—being the transaction or transformation of one form of capital

---

[4] The figures referred to here from the 2002 Review are: $4.7 million were spent on investments to $1.4 million on education, training and business development.

[5] Another reason for the increased long term investment, to be examined below, was that insufficient proposals for business enterprises were put forward in the first five years. As the Review makes clear, this is indicative both of the lack of support for developing such proposals and the perceived (and actual) constraints on what constitutes an acceptable proposal.

[6] The Graham (Polly) Farmer Foundation is a philanthropic organisation that provides after school mentoring and homework facilities for Aboriginal high school students in towns in the Pilbara region and other supportive roles. Rio Tinto is also a partner through their Futures Fund.

into another.[7] In this case a more significant proportion of the financial capital from the YLUA could be available to be transformed into human, social and cultural capital.[8] Moving beyond the focus on the agreement as primarily an economic asset is a step toward this approach.

It seems that this is the approach being increasingly adopted by Gumala, as is possible within the confines of the YLUA. Gumala grappled with the findings of not only the first five-year Review (Hoffmeister 2002), but now the second Review which was finalised in August 2008.

Gumala Aboriginal Corporation, the Trust and Business arms

'Gumala' is an Aboriginal word in the Bunjima Language meaning 'all together'. This ideology was important in binding three language groups—Bunjima, Innawonga and Niaparli (who maintain native title in the region of the Yandi mine and associated infrastructure)—together in the YLUA.[9] However, holding them together has been a major challenge with ongoing tension between elements of each group, not least because there now are over 750 Aboriginal beneficiaries, party to the YLUA. Some of these tensions, recent and historical, will be discussed below.

There are four bodies that operate under the YLUA: Gumala Aboriginal Corporation based in Tom Price, its business arm GEPL, the Gumala General Foundation, and Gumala Investments Pty Ltd (GIPL). GIPL was established to act as the Trustee of the General Foundation, which receives payments of between $2 million and $5 million per annum from the YLUA.[10] Importantly, the Trustee is independent of Gumala and has ultimate decision-making powers in all matters relating to the Foundation. A Foundation for the elderly and infirm was

---

[7] Pearson and Kostakidis-Lianos (2004) argue that Aboriginal land-holding structures and property rights, such as native title, are not readily fungible into economic assets. In fact they state that land rights are 'dead capital' because they cannot be leveraged to create capital. They give the example of setting up a private enterprise on Aboriginal Land as a 'virtual impossibility' (Pearson and Kostakidis-Lianos 2004: 2). However, it must be pointed out here that the agreements under consideration in this paper were negotiated under the *Native Title Act 1993* —hence leverage was gained through Aboriginal property rights and the outcome is an agreement. Such an agreement is a very significant economic asset. Likewise, under the *Aboriginal Land Rights (Northern Territory) Act 1976,* the right of consent provisions also provide significant economic leverage.

[8] There is a vast literature on the concept of social capital with the World Bank, in particular, bringing the concept into prominence with their interest in understanding the local impediments to development in considering the social as opposed to the economic face of adjustment (Woolcock and Narayan 2000). Critiques of the concept (Fine 2003; Hunter 2004; Renzio 2000; also Woolcock 1998 before joining the World Bank) have noted that there is a marked neglect in how social capital is actually created, that it is a catch all category designed to capture any asset that does not fall under the conventional categories natural, physical and human. These authors also focus on the definitional chaos of the concept.

[9] The orthography used here for these language groups follows the Gumala spellings. Note, however, that there are a range of alternatives. The orthography preferred by the Wangka Maya Pilbara Aboriginal language centre based in Port Hedland is as follows: Banyjima, Yinhawangka and Nyiyaparli.

[10] This annual amount is currently closer to $5.5 million (Darren Inji, pers. comm. June 2008).

established to provide benefits to those so defined for the first five years. It no longer operates. Apart from managing and maintaining the capital base of the General Foundation, GIPL considers project funding requests from Gumala. Gumala is charged with consulting members of the beneficiary group, and developing, researching and preparing proposals for investments and community projects, as the on the ground Indigenous organisation. Gumala is thus the manager of GIPL and the sole shareholder of GIPL as Trustee. GIPL comprises six directors, three directors who are independent and three directors who are traditional owners as representatives from the three language groups.

Many of the issues in the following discussion are drawn from the 2002 Review of the Trusts (also known as Foundations) and GIPL, along with interviews conducted over 2003–04, with members of Gumala across the Pilbara and, more recently, telephone interviews with some of these same people. Although the Review was not of Gumala or its business arm GEPL, a number of the Review findings directly engage with issues associated with their operations. Many of the issues of concern raised in the 2002 Review remain pertinent, although a number of the recommendations have been acted upon. There are several apparent reasons for the continuing relevance of the 2002 Review findings. One of these was described as the lack of communication between the two key bodies—Gumala and GIPL—and between these bodies and their members. Indeed, in early to mid 2007 Gumala attempted to liquidate GIPL to, as one commentator noted, 'get their hands on the booty' (Bill Day, pers. comm. May 2007). At June 2007 this 'booty' was worth approximately $20 million (Siopsis J 2007). GIPL Trustees and other supporters applied for and won an injunction against their liquidation and, to avoid protracted legal costs, a mediated settlement was reached. This issue, to be discussed further below, is raised here to situate some of the tensions between these organisations and to place at the forefront this issue of the contested access to the benefit stream.

The Review found that most available income was invested and that considerably less was spent on program areas, especially education, training and business development.[11] While at that time GIPL and the Gumala CEO stated that there had not been enough proposals, members said that they found it hard to put proposals forward and to understand the approval process. It seems that the most 'business ready' and 'job ready' tended to be most advantaged and the approach continues to be that business proposals must have a prospect of making a profit. Gumala business development funding had done little for groups who were not previously business ready. A key recommendation was to establish a

---

[11] Indeed, the tensions inherent in this pattern of investment and expenditure; getting the balance right, have a long history. These issues are not unique to this Agreement, but rather were apparent in the earliest royalty associations and trusts set up to disburse mining monies in the Northern Territory (see Altman 1983). What is perhaps different here is both the scale of the Agreement and the Indigenous expectation.

business development assistance scheme within 12 months of the Review to be situated within Gumala (Hoffmeister 2002). A Business Development Officer was employed for a short period, but at that stage it was not successful and was discontinued. A Business Loans Guarantee Program has also been operating, but with little success. In late 2008, the Business Loans Guarantee Program was in the process of being transferred from GIPL to Gumala in order to facilitate greater utilisation. Since the first Review, members' businesses which have been supported include a cattle station, an earthmoving enterprise, a cultural awareness business, a fabric and garment design enterprise and a bush products enterprise (see Lola Young's case study below).

Exactly what sort of 'business development' was to be supported was found to be a major area of uncertainty and dissatisfaction for Gumala members interviewed for the Review. The Gumala newsletter at that time highlighted that, 'All funds from the General Foundation must go towards community development projects and NOT INDIVIDUALS' and further that, 'All language groups are encouraged to submit proposals, as outlined in their community action plans' (*Gumala News*, December 2001, capitals in original). Without having the benefit of sighting any 'community action plans', it seems that members would be justified in their confusion about what constitutes a successful proposal. It was noted that individual and family business development proposals were rejected when they were put forward. In the 2002 Review the Trustee indicated that a cautious approach was taken in only supporting projects with a high probability of success. Such projects were expected to have a 'community benefit'. This raises the question of how benefits to individuals and families are distinguished from broader community benefits. This distinction would be especially blurry where these individuals and families are living on homelands—that is, small communities. At least two family/individual business have been supported.

In interviews held with Gumala members, the issue of not funding individual or family enterprise was seen as not only extremely limiting, but as a potential liability. Interviewees' understanding was, that an entrepreneurial individual could only be allowed to work with others in the local homeland who may or may not be interested in the project, despite others, such as non-Indigenous people, being able to input greater skills. A general comment was that an entrepreneurial individual did not want to have to rely solely on their family or those in the local homeland for success. This is consistent with Gumala members noting in the 2002 Review, that supporting individuals in small business is a key to securing economic independence. Likewise, it is arguable that encouraging such individuals as role models may have wider ramifications. This is the approach adopted by the United Nations Development Program, discussed earlier, and accepted internationally as a means of growing employment possibilities within a community.

Cash payments to beneficiaries cannot be granted under the Trust structures. This is for some, who are critical of this approach, as much a political and rights issue: believing that they have a right to manage their own finances, cash payments or otherwise, like other Australians. Many of the Gumala members interviewed found the lack of choice patronising, and there was little understanding that the charitable status of the Trusts was a reason for not granting such payments. Gumala members were well aware that the neighboring IBN (Innawonga, Banyjima, and Niapali) Agreement with BHP Billiton, also of comparative scope and majority overlapping membership, does provide limited annual cash payments to members. If membership of Gumala is viewed as part of a mosaic of LUAs in the region, this feature of the IBN Agreement could be understood as assuaging, to some extent, the call by those individuals who are members of both.[12]

The 'community development' that is occurring is homeland (outstation) infrastructure development and support for associated cultural activities. Six homelands in particular, located mainly in the mine hinterland, are the recipients of these monies. Projects supported include roadworks, tractors and sheds, medical rooms, a community bus and a breakfast and homework centre, and one homeland received funding for essential services. This activity—supporting small dispersed settlements on land where particular families have rights by customary law—seems to be almost universally regarded as adding value to people's lives and was widely discussed by Gumala members as positive. The 'priorities for the future' listed as part of the 2002 Review noted an emphasis on returning to country with the assistance of community development projects and the preservation of law and culture; but also included business development loans for individuals. Accordingly, community development was not understood by Gumala members as excluding small scale business development.

There is the risk that the money being spent by Gumala on essential services could have been spent instead on fostering small scale business. As O'Faircheallaigh (2004a: 43) suggests, 'If mining payments are used to pay for basic social services [that are citizenship rights] then opportunity'... to utilise a significant economic asset cannot be utilised to overcome economic disadvantage. A case can be made that the development of these homelands has been an example of 'substitution funding', whereby the expenditure from mining

---

[12] Calls by many Gumala members for individual entitlements in the form of cash payments can also be considered as another form of entrepreneurialism, taking into account that a fundamental feature of entrepreneurialism is an individual, as opposed to a collective, approach to engaging in financial opportunities. Nevertheless, I acknowledge that such compensatory or cash payments may be regarded as passive, whereas entrepreneurial activity is necessarily active. However, for my purposes here, the immediate expenditure issue needs to be understood in light of the framework of constraint that the YLUA places on individual capacity to choose, as both a political and social right. Therefore, it can be considered as falling under the Aboriginal Canadian definition of entrepreneurialism (Peredo et al. 2004), as a broader definition than that offered by Hindle and Lansdowne (2005) and Foley (2006).

payments has substituted for government funds that were spent elsewhere. The result is no net increase in spending on services in these communities (see O'Faircheallaigh 2004a: 43). This issue of 'substitution funding', potentially jeopardising access to commonwealth or state-funded programs because of mining agreement monies, is not a new one. Pre-native title, this risk was most notable where significant financial benefits were negotiated in the context of land rights legislation in remote areas (see Altman 1983: 112, 1985a). Gumala was also not unaware of this risk and in 2004 the then CEO and Senior Project Officer both spoke of the value of projects that leveraged government resources. Such projects included the medical rooms at one homeland that were built by YLUA funds, as the government would not provide capital infrastructure funds, but would provide recurrent resources for staffing; and the homework centre at a homeland where the infrastructure was supplied by Gumala, but the computers and network were subsidised by the federal government 'networking the nation' program. The risk of substitution funding remains significant and requires ongoing strategic management (see Altman, Chapter 2; Levitus, Chapter 4). In view of this, Gumala and GEPL employed a Grants Officer in August 2008 to ensure that they leverage grant money and sourced available opportunities.

As suggested by the discussion above, the 2002 Review found general dissatisfaction about the Trustee and the GIPL manager handling funds for education, training, and business development. A number of the criticisms to emerge from the 2002 Review revolved around the perception that GIPL was constraining opportunities for accessing the YLUA monies. Members were concerned that the Trustees were too far away, several resided in Perth, and knew little of their concerns and needs. In an earlier paper I have indicated that some Gumala members perceived 'that they have insufficient control over how the money in the Trust is distributed' (Holcombe 2004a: 9–10). Part of the reason for this is the decision-making structure of GIPL. There are three non-Indigenous and three Indigenous representatives (one from each language group) on the Board of Trustees, on three-year terms and meeting twice annually. In 2004, however, they were only having one face-to-face annual general meeting, to save costs. This sense of disconnection from the decision-making process may be inevitable, however lack of transparency in the decision-making process and slowness of decision-making were cited as significant in the 2002 Review.

The 2002 Review found a need for more cooperation between the Boards of GIPL and GEPL (as the Gumala Enterprise arm), and for better information dissemination, communication and consultation by the GIPL Manager and Trustee with Gumala members and Gumala officers. The issue of GIPL having ultimate decision making power over Trust funds while Gumala as manager could only pass on funding proposals, did not sit well with some Gumala members and especially the then CEO. This tension between Gumala and GIPL came to a head in 2007 when Gumala attempted to sack the GIPL Trustees and appoint a liquidator who would enable compliant trustees to then transfer the remaining $40 million investment fund to Gumala. For its part, GIPL maintained that it had concerns related to the manner in which financial controls were exercised and the lack of proposals being put forward by Gumala (Siopsis J 2007). GIPL lodged an injunction against its liquidation on the basis that there had not been a full vote of members and Gumala (as manager of the Trust) had breached its fiduciary duties to the beneficiaries in its lack of openness. The court ordered a mediated settlement and shortly thereafter the Gumala CEO, who had held the position for approximately five years, resigned. A new CEO was appointed in January 2008 and a restructured Gumala committee was established.[13]

Finally, a number of the 2002 Review recommendations have been acted upon. For instance, the Review recommended that children of Gumala members be supported through the provision of scholarships for secondary school, higher education, and for post-school vocational training. A number of children now attend Perth secondary schools on scholarships. In 2006, the business arm of Gumala, GEPL, invested in the Karijini Eco Retreat tourism venture, where there are currently 10 local Indigenous people employed over a nine month tourist season. This approach to business development acknowledges that diversification beyond the mining industry is a strategy to distribute risk.

In early 2008,[14] with a new Governing Committee and a new Chief Executive, Gumala made major changes in its structure by increasing numbers of staff, creating an integrated relationship with GEPL, and implementing fundamental changes in relation to business development. These include the expansion of existing corporate businesses and the development of others, such as a major accommodation project in Paraburdoo for Rio Tinto staff, contractors and tourists and a proposal for a large business complex in the centre of Tom Price with shops, offices, training facilities, accommodation and new corporate head quarters for Gumala. A new strategic plan is also being developed so that GEPL can seize upon the opportunities created with the Pilbara mining boom.

To cater for individual circumstances a Members' Support Unit has also recently been established with extra staff to deliver direct benefits through various programs. These include sport and recreation, financial support for funerals and headstones, emergency assistance, a pensioner program, critically ill patient support, health and wellbeing programs, dental and ancillary health, computers

---

[13] This committee restructure was also partly due to an Australia-wide Office of the Registrar of Indigenous Corporations development that reduced the size of committees for Aboriginal corporations by half. Gumala had previously required 18 elected members, six from each language group. Now the requirements are three members of each language group, giving a maximum of nine. Although smaller boards/committee may be more efficient and cost effective, this may be at the expense of transparency, and real and perceived exclusivity.

[14] The detail in this paragraph and the following paragraph was provided by Gumala (D. Gordon, pers. comm. 10 October 2008).

for students, a scholarship scheme (discussed earlier), and home loans. Seven trainees have been employed at the Gumala office, while GEPL is also expanding its Indigenous workforce.

Individual Indigenous engagement

To illustrate something of the diversity of Indigenous engagement with mining interests, biographical details of several individuals are discussed below. This material covers not only consideration of some small and large scale business activity, but also process-oriented engagement that speaks to a political and cultural agenda, rather than solely financial motivation.

Case 1: PN

PN is an articulate young to middle aged Niaparli woman who resides on her family outstation in the Hamersley Ranges. In 2004 she was community President. At that time she was a member of seven committees, boards and working groups, including Gumala, GEPL, the Pilbara Native Title Service (PNTS) Regional Committee, the Yamatji Marlpa Barna Baba Maaja Aboriginal Corporation (YMBBMAC) Governing Committee,[15] the Niaparli (native title) working group, and the Rio Tinto Central Negotiating Committee.[16] Her membership of these organisations, involves a strong element of 'keeping a finger on the pulse' of the regional socioeconomic politics. In a 2005 PNTS press release, PN spoke strongly about Rio Tinto's need to maintain standards of corporate social responsibility and to negotiate retrospectively, given the expansion of mines that were developed prior to the *Native Title Act 1993* without traditional owner approval.[17] The level of commitment to monitoring the massive industry presence is apparent.

At the time of interviewing PN and compiling the above membership list, it seemed one, extreme, end of a spectrum of engagement—the pointy end of the process side of the mining boom. However, this type of engagement seems to suit many Indigenous people and is not necessarily uncommon. What is perhaps striking about PN's membership commitments is the level of activity. While some individuals feel a clear need to engage politically with such a powerful industry group, there is a flexibility to attending meetings which caters to other

---

[15] This governing committee oversights the policy direction of the YMBBMAC. This is the organisation that acts as the native title representative body for the Pilbara region, as well as the coastal region around Geraldton to the south west. It incorporates the Yamatji Land and Sea Council and the PNTS.

[16] The Central Negotiating Committee was formed in 2003 to engage collectively with Rio Tinto Iron Ore (RTIO) in developing a regional framework through which to negotiate a coordinated approach to the ongoing expansion of the industry and the new and evolving agreements. In 2006, the Central Negotiating Committee developed into a private company owned and controlled by 10 Pilbara traditional owner (language) groups. With an office in Roebourne, chaired by Slim Parker, the Marnda Mia Central Negotiating Committee 'aims to build coordinated, institutionalised capacity for local Indigenous families and groups and provide a strong local voice' (see Rio Tinto media release, 26 September 2007).

[17] 'Pilbara Aboriginal meeting condemns Rio Tinto', PNTS media release, 31 May 2005.

Indigenous priorities such as funerals and ceremonies. If one is unable to attend, for whatever reason, a proxy can usually be appointed. Attending regular meetings in regional towns and cities also caters to patterns of mobility, suiting the mix of responsibilities that individuals have to both families and localities.

Sitting fees of $500 per day for some of these committees makes this permanent part-time range of commitments to various organisations sustainable for participants (although presumably each person has to manage potentially complex tax implications). Some Gumala members have argued that committee memberships can be understood as part of a 'mining welfare economy' that tempts individuals to remain available at the expense of gaining full-time employment. Other permanent part-time work also includes heritage surveys and the monitoring of infrastructure works: the expansion for iron ore extraction requires almost on-going survey work. In relation to Rio Tinto Iron Ore (RTIO) operations,[18] in 2006 alone in the Pilbara there were 2578 Aboriginal consultant days over 96 surveys, with further expansion expected in the following years (RTIO 2006: 32).

Case 2: Lola Young

Lola Young, aged in her 60s, lives in Tom Price. Young founded Wakuthuni, probably the first homeland in the Hamersley ranges, in 1990. An excision was negotiated with the Rocklea pastoral lease owners at the time with the assistance of the Aboriginal and Torres Strait Islander Commission (ATSIC); the lease is currently owned by RTIO. Young was one of the first members of Gumala and during 2004 was also a member of the Innawonga working group. Last year her biography was published with the assistance of Pilbara Iron (part of RTIO) (Young 2007).

As a knowledgeable Innawonga elder, Young has a high regard for the cultural value of land and maintaining attachment to it. When referring to the best methods for teaching children she has stated that 'every long weekend we need to get and teach them. If we teach them from outside of our land we get no strong inside feeling from them. You can feel it really strong when you are talking from your own land' (in Olive 1997: 99). Young has been involved with establishing two local businesses which clearly speak to this priority of cultural maintenance: Wanu Wanu and Ngumee-Ngu. The Wanu Wanu Aboriginal Corporation was established in 1997 with support from Hamerlsey Iron's (now Pilbara Iron) Aboriginal Training and Liaison Unit as a cross-cultural training business: Hamerlsey Iron employees could stay overnight in the Wakuthuni homeland as part of a suite of cultural awareness training. It was to extend to cultural and eco-tourism, but was de-registered in 2004. At the time, there was discussion

---

[18] The RTIO operations—through Pilbara Iron— operates seven mines (including Yandicoogina) and associated infrastructure.

about linking the proposed cultural and eco-tourism project in the neighboring Karijini National Park (that Gumala was planning on supporting) to Young's Wakuthuni walking tour.

The Ngumee-Ngu Aboriginal Corporation (the name of which derives from Young's 'bush name'—Ngamingu) was established in 2002, with the support of Gumala (Young 2007: 84). The objects of this corporation are listed as: 'to become self sufficient; to care for the country, the corporation and the people; to build homes within the homeland and to provide economic; social and cultural services to the community' (Office of the Registrar of Indigenous Corporations 2008). Young's biography features local knowledge about flora and its medicinal qualities, and the business includes the manufacture and sale of bush medicines based on this knowledge. The products have been sold at the annual Mount Nameless festival in Tom Price. These businesses are concerned with social and cultural cohesion and maintenance, rather than being driven principally by financial motives—although that is an aspect. Likewise, inter-generational transfer of knowledge is cited by Young as important—she works with her grandchildren on the production of the bush medicines (Young 2007: 83). The biography contains a CD of six songs performed by Young; songs learnt from her parents and from the spirits of her country (Young 2007: 160).

Case 3: ID

ID is a young to middle aged man who resided on his family homeland in the Tom Price region during 2003 and 2004. At that time he was the Secretary of the Gumala Committee, the Secretary of the GEPL Board, a member of the Innawonga Working Group, and a committee member of the Pilbara Fund (a program sponsored by the Pilbara Development Commission in Western Australia).[19] He had early involvement as an activist against the Rio Tinto Marandoo mine development into the Karijini National Park[20] and was part of the early push toward developing cultural and eco-tourism in the Park. To this end ID was actively involved in establishing the Karijini Aboriginal Corporation in 1991 to facilitate a local Indigenous tourism operation in the Park. The first object of this Corporation was listed as, 'to support the social development of its members in all ways', followed by 'to help bring about the self-support of

---

[19] 'The Premier announced the $20 million Pilbara Fund on 26 July 2004 to accelerate investment in the community and Government infrastructure throughout the Pilbara, particularly in the areas of health, education, recreation, culture and Government housing. The primary objective of the Pilbara Fund is to add to the welfare of all people of the Pilbara and to make the region a better place to live and work. It is intended that the Fund will facilitate the social and economic development of the Pilbara through funding projects that will enhance the long term sustainable future of the region' (Pilbara Development Commission 2008).

[20] For detail about this dispute see 'On the fast track to a dispute over Marandoo', M. Steketee, *Sydney Morning Herald*, 5 November 1991, p. 10; 'Showdown at Marandoo', M. Stevens, *Business Review Weekly*, 6 December 1991, p. 76.

its members by the development of economic projects and industries' (Office of the Registrar of Indigenous Corporations 1990).

The Karijini Aboriginal Corporation was de-registered in 2004. However, these ideals have flowed into the development (by Gumala's business arm, GEPL) of the Karijini eco-tourism enterprise outlined above. As a GEPL Board member, ID was active in ensuring that tourist infrastructure be developed and formal arrangements be finalised for the camp grounds all under Indigenous control. As he noted himself, '…as a valuable tourist industry … we can show off the Aboriginal culture with the aim of preservation and protection' (in Olive 1997: 205). Land related enterprises have been a signature of ID's business pursuits. In 2004, his small business sold up to 300–400 kangaroo tails a month. Under the YLUA, Rocklea Pastoral Lease (owned by RTIO) was to be returned to Gumala members. This 'handover' had not occurred in 2004, and I am uncertain as to the current situation. Nevertheless, given the early Indigenous engagement with pastoralism in this region there is a certain romance in returning to this era of stock work, now dramatically overshadowed by the mining industry, which now owns the majority of Pastoral leases in the Pilbara region. Indeed, most Indigenous people under the age of 50 would not have been exposed to the culture of station life (see Holcombe 2006: 81).[21]

ID developed a business proposal to utilise a herd of cattle already marked for slaughter (known as 'killer' cattle) for a local meat supply, requiring associated yards and fencing between two of the homelands near Tom Price. This project was not necessarily intended to make a profit and was reliant on the Community Development Employment Program (CDEP) for part time wages. ID indicated at that time that the project was not well received by Gumala. It would seem the proposed business had potential to grow, would assist in training young men for stock work, and offer a steady supply of meat to community residents with limited resources. This sort of enterprise is driven by a desire to return to the land, and to develop an alternative informal or domestic economy.

ID could be described as a political and cultural entrepreneur: as a strong supporter of a unified Gumala he was closely involved in the injunction against Gumala opposing GIPL's liquidation. As ID has noted, 'I have always been a fierce supporter of land rights ever since I knew what land right's was about (in Olive 1997: 204).

At the time of the field research, the three individuals above had chosen not to be involved with the IBN Corporation, even though there were clear financial incentives to do so. This Corporation covers the BHP Billiton mine (known as 'Area C') which is an immediate neighbor to the RTIO Yandi mine. Without

---

[21] When compulsory schooling for Indigenous children was introduced in Western Australia in the late 1960s and early 1970s there was a major relocation of Indigenous workers in the pastoral industry to the towns (see Holcombe 2006: 81).

wishing to unduly idealise the reality, this speaks to the deeper issue of group identity and the fault lines within which the regional polity is constructed. The membership rights of these three individuals to the IBN Corporation would be unquestioned and immediate, given their relationship to the native title groups through which the IBN Corporation is constituted—the Innawonga, Banyjima, and Niapali.[22] However, a brief consideration of the way in which the IBN Corporation was established may explain their decisions.

When the LUA was being negotiated for the Area C mine, Gumala was well established and there was an expectation by many Gumala members (based on my interviews conducted in 2003–04) that Gumala would be the organisation that would negotiate the Agreement, which would aid its growth as a regional Indigenous organisation. Certainly there was, theoretically, 100 per cent overlap in membership of the two Agreement groups.

The concept of an overarching organisation that centralises, and thus standardises, negotiations was in some ways a forerunner to the Central Negotiating Committee that has now become the Marnda Mia Central Negotiating Committee (see earlier footnote). However, the inaugural Chairman of Gumala left when his position was not renewed, and he pursued the role of CEO of the new IBN Corporation. The significant contestation between the two organisations indicates the competition for the hearts and minds of the membership, as well as competition for Indigenous workers—both organisations have Indigenous contracting services (Gumala runs Gumala Contracting and IBN Corporation runs Indigenous Mining Services) that compete for staff.

The decisions of the three individuals discussed above are not principally motivated by financial gain; rather, their agendas articulate closely to an entrepreneurial paradigm that is underwritten by cultural and political values. As ID has stated, 'I am a strong believer in Aboriginal culture. What a lot of people don't realise is that Aboriginal culture is moving fast and adapting' (in Olive 1997: 203). These three short biographies, and indeed this quote, seemed to me to typify the types and levels of engagement that members of Gumala have with the Agreement. That is of those individuals, who choose to engage, maintaining autonomy in a manner that resonates with an Aboriginal agenda remains a significant prerogative.

Case 4: SC

The first Chairman of Gumala, SC is a more mainstream or classic entrepreneur. SC trained as a boiler maker and had worked for BHP Billiton for nine years by the time he became Gumala Chair. When he was not reappointed Gumala Chair for a second term, he developed the IBN (Innawonga, Banyjima, and Niapali)

---

[22] The IBN Corporation was an outcome of a Land Use Agreement (LUA) with BHP Billiton, as Gumala was with RTIO.

Corporation to manage the Agreement for the neighboring BHP Billiton mine.[23] This dynamic fissure and fusion of organisations, illustrated in the de-registration of several of the corporations discussed above, is a backdrop to the pattern of leadership. New organisations are established as existing ones cease to serve as the vehicle to progress the founder's aims and ideologies. As I have noted elsewhere, the negotiation of the YLUA was described as 'learning curve' for SC (Holcombe 2004a: 12). When an opportunity arose to develop another organisation, it was taken and the IBN Corporation was developed.

The structure of this Corporation differed significantly to that of Gumala and it is clear that the inaugural CEO transferred some lessons from his experience with the YLUA structures. Like the YLUA, two Trust funds were established under the IBN Corporation. However, unlike the Yandi Trusts, one of the IBN Corporation Trusts, the Financial Assistance Trust, did allow cash payments to members, as it was designated as non-charitable. As noted above, the issue of cash payments was significant for some Gumala members (see also Holcombe 2004a: 12–13).

Another significant difference in the IBN Corporation structure was the incorporation of four discrete bodies representing the language groups of the native title claim.[24] Each group has their own administrative support and funding stream. With parallels to the issue of cash, the independence of each language group has been an ongoing source of tension for some groups within Gumala. The 2002 Review found that amongst each language group some individuals sought to devolve the current structure to language group corporations.[25] However, agreement could not be reached so no action was taken. Some of this dissatisfaction lay behind the Gumala attempt to liquidate GIPL in 2007, discussed above.

A third significant difference between Gumala and the more recent IBN Corporation was the centralised authority structure that was developed by SC, as the inaugural CEO. Instead of two bodies—one that manages the Trusts (GIPL) and another operating arm (Gumala)—the IBN Corporation is structured so the CEO and the Board of Directors have oversight over both the IBN Corporation as Trustee and the operating arm (the contracting business Indigenous Mining Services). There is no separation of powers between the advisory body and the decision making body. This has been a critical issue for the IBN Corporation and member concerns about accountability and transparency.

---

[23] Note these language spellings are those used by the Corporation.

[24] These are the Banyjima, Niapali, Miluranpa Banyjima and the Minadhu Innawonga groups—the two different Banyjima groups are also known as the Top End and Bottom End Bayjima groups. Note that these are the spellings used by the IBN Corporation.

[25] However, there was also awareness amongst members of the significant cost implications of this administrative duplication.

The four individuals in the case studies briefly discussed here are not necessarily doing anything extra-ordinary. Rather, within the limitations imposed they are negotiating their needs and in some cases pushing the boundaries.

Conclusion

Focusing on the issue of entrepreneurialism brings the tension between collective or communal rights and the rights of the individual into stark relief, although this chapter has not explored this tension in Aboriginal political process. Reconciling these apparently opposing Aboriginal values and practices is a key challenge in utilising the mainstream opportunities that the agreements offer (see Martin 1995, 2001). Mining agreements could be about offering choice and acknowledging the diversity of expectation within the Aboriginal stakeholder group, rather than operating as an experiment in social engineering. In the context of the LUAs discussed here, Indigenous entrepreneurialism is not just about engaging in the 'real economy', but also about enabling and encouraging individuals in all their heterogeneity to pursue a diverse economy. As an Aboriginal 'beneficiary' of the YLUA, Keith Lethbridge, suggested 'company structures … [should not only] be to generate money' (Ethical Investor 2004: 33), and it may be added, not just to be directly supportive of the mining industry (see Scambary 2007, Chapter 8).

The 2002 Review into GIPL and the Trusts showed that where Aboriginal businesses are supported, they have had to be low risk and show a direct 'community benefit' (Hoffmeister 2002). Such businesses tend to be in industries that service the mine economy (see Hamersley Iron 2000). Rio Tinto's Aboriginal Training and Liaison Unit provides examples of the 'diversity and scope of business opportunities that are made possible by the Hamersley program', all are supportive of the industry. Two businesses—Wanu Wanu (discussed above) and Ngurra Wangkamagayi—run cross-cultural training courses to Hamersley Iron (now Pilbara Iron) and other companies; Ieramugadu gardening services provide contracting maintenance to Hamersley's port operations in Dampier; GEPL which operate a contracting arm (earth moving); and Gumala Eurest which operates camp accommodation and associated services.

Although clearly this is where most business opportunity exists in this region, such mainstream 'opportunity' is only taken up by 30 per cent of Indigenous adults (Taylor and Scambary 2005: 28). This suggests that looking beyond the life of the mining industry is a fanciful exercise when the pressing issue now is 'how can *more* Aboriginal people benefit from an LUA in terms that suit them rather than the industry?' As indicated above, these terms may be less driven by economic imperatives, than by cultural and political ones. Some realism is needed here in regard to diversifying the range of benefits that can be gained from LUAs, as it seems that the majority of those imputed to be 'beneficiaries' in fact benefit very little, if at all, unless they are directly employed by the Company or sit on the range of committees to undertake the process work.

While an important component of LUAs is to enable engagement with the mainstream economy through employment and training programs, choice should not be limited to this. Nevertheless, such mainstream opportunities need to be more inclusive of the disadvantaged and disenfranchised. More of the agreement capital could be allocated to building diverse forms of capital—human, social and cultural—to encourage entrepreneurialism in its many forms. The issue of whether business development proposals are specifically driven by the market should not be the only consideration in approving proposals, and nor should the size of the benefit group or community. Importantly, establishing a business development assistance scheme emerged as a means to enable a greater range of individuals, not only those that are business ready, to access a greater range of opportunity. Because the success of an individual is not an isolated achievement, the provision of scholarships to schools of the parents' choice, for instance, is a crucial element of this. It seems that the Rio Tinto WA Future Fund with the Graham (Polly) Farmer Foundation, and Gumala's new program, offers some scope here.

Limitations on choice create social pressures and fissures which highlight contestation over value and need, and Aboriginal notions of success. To capture these values, a broader consideration of 'opportunity' needs to be canvassed. This concept of opportunity, perhaps like 'capacity building', is currently based on channeling Aboriginal interests toward engagement with the mainstream and formal economies, primarily through the uptake of employment in the mine economy. When individuals are targeted by agreement programs, it tends to be in terms of specific training for skills appropriate for employment in the mining sector, and aiming at full-time employment. Engaging with this employment regime or meeting the guidelines for businesses funded under the trust structures is premised on 'opportunity', which refers to the opportunity to change—that is, to change value systems, if one is not already 'business ready' for instance (Holcombe 2006).

This issue of what might constitute Indigenous entrepreneurialism is perhaps nascent in remote Australia, as is the notion of community economies in these regions. Nonetheless, it is clear that an emergent hybrid economy is evolving in the mine hinterland region of the Pilbara. By leveraging the right to negotiate under the native title act and striking land use agreements, individuals are less focused on harvesting game (the customary economy) and more driven to harvesting heritage, through site clearances for mine works, and the development of homelands. While not every Gumala member seeks to reside on a homeland, or indeed has the customary right to establish one, neither do all Gumala members chose to, or are enabled to, work in the mining industry. Getting the balance right in the YLUA between catering for the range of expectations of outcomes is of course unique to it, as it is to other agreements. The purpose of this chapter has been to re-direct or refocus attention onto individuals and the ways in which they are able to articulate with the Agreement through the transformation of economic capital into social, cultural and political capital. The development of the Marnda Mia Central Negotiating Committee is a powerful example of the Agreement acting as one lever, of several, to build political capital. Likewise, local economies which revolve around cultural and eco-tourism, the manufacture of bush products and so on, also deserve space and should not be overlooked, even in the context of a regional mining boom. I return to an earlier quote: 'if the money is not being spent, it is not doing good'. This chapter has hopefully opened up the field of discussion in this space about what 'good' might mean as it is applied to a remote nascent entrepreneurialism.

8. Mining agreements, development, aspirations, and livelihoods

Benedict Scambary

In a time of global economic and climate uncertainty the expansion of resource exploitation projects in Australia is unprecedented. The consequent value of the minerals sector to Australian prosperity is in stark contrast to the economic poverty experienced by many Indigenous Australians, particularly those residing in mine hinterlands. This contrast is evident despite the existence of beneficial agreements between Indigenous groups and the mining industry, and in some cases the state, concerning the very mining that is generating such extraordinary profit. Indigenous poverty is only minimally ameliorated by such agreements (O'Faircheallaigh 2000, 2003a, 2004a, 2006; Taylor 2004a; Taylor and Bell 2001; Taylor and Scambary 2005). This chapter draws on doctoral research of three agreements to highlight that poor agreement outcomes arise from the limitations that agreements impose on Indigenous livelihoods and aspirations (Scambary 2007). The agreements considered by this research are the Ranger Uranium Mine Agreement in the Kakadu Region of the Northern Territory, the Yandicoogina Land Use Agreement (YLUA) in the Central Pilbara of Western Australia, and the Gulf Communities Agreement (GCA) in the southern Gulf of Carpentaria in Queensland. All three agreements are considered best practice by the mining industry, the state and select Indigenous leaders, for their perceived capacity to deliver substantial and sustainable benefits to Indigenous people. However a combination of the scale of Indigenous disadvantage and the mainstream development parameters of the agreements themselves limit the attainment of sustainable outcomes for Indigenous people associated with all three agreements. This chapter argues that a fundamental limitation of these mining agreements is their incapacity to engage with and augment the diverse livelihood objectives of Indigenous people. This motivates ambivalent responses to mining on the part of Indigenous people. The invisibility of Indigenous cultural dispositions is further compounded by a growing policy emphasis on mainstream economic engagement, which entails many negative assumptions about Indigenous people and their capacity for economic engagement. Such assumptions are in tacit opposition to Indigenous cultural dispositions that ultimately underlie notions of identity and a claim to rights on the basis of cultural difference.

Indigenous people's experiences of initiatives promoting 'sustainability' in the context of mining agreements are primarily associated with royalty and compensatory payments directed at attaining community benefit. In modern mining agreements 'community benefit' packages broadly entail employment and training programs, business enterprise development programs, payment of limited cash compensation, and heritage protection. Successes and failures within these realms are *ad hoc* within each of the agreements, with all three falling well short of their objectives to overcome Indigenous disadvantage via the creation of economic opportunity (see below). The reasons for this are numerous and complex, but include:

- the challenge of reconciling mainstream economic development initiatives (that seek outcomes almost exclusively in terms of economic engagement associated with the local mine economy) and Indigenous cultural prerogatives (that arise from, and construct, personal and group identity and are often expressed as livelihood aspirations)
- the level of accord between defined agreement beneficiaries and local Indigenous conceptions of relatedness
- the ability of Indigenous organisations arising from the agreements to represent the diversity of their memberships
- the effects of statutory and agreement defined conditions on the flow of benefits to intended beneficiaries
- the tendency of the state to retreat as a service provider with the arrival of private capital, and
- the extent of Indigenous autonomy over agreement benefits.

The Agreements

The structures of the Ranger Mining Agreement, the YLUA, and the GCA are complex and define the space for Indigenous 'productive activity' within their contexts. Generally all three agreements provide for compensatory and 'community benefit' packages—trust funds and royalty equivalent payments, programs of employment and training, business development, education, and cultural heritage protection. Whilst all three agreements fall within the mining industry's sustainable development approach, critical differences emerge between the agreements that include the role of land councils and native title representative bodies (NTRBs), the intended purpose of funds arising from agreements, and issues of governance associated with agreement structures.

The Ranger Uranium Mine is located on Mirrar Gunjeihmi country within Kakau National Park in the Northern Territory. Approximately 50 per cent of Mirrar Gunjeihmi country is encompassed by the Ranger and nearby Jabiluka uranium mining lease. The Ranger Mining Agreement, pursuant to s.44 of the *Aboriginal Land Rights (Northern Territory) Act 1976* (Cwlth) (ALRA), is between the Northern Land Council on behalf of traditional owners and the Commonwealth. Analysis of the Ranger mine's nearly 30 years of history is instructive for understanding relationships between Indigenous people and the modern mining industry in Australia more broadly. Many conclusions can be draw from the impacts on the local Indigenous polity from the construction of a dedicated mining town (Jabiru), Indigenous employment and training schemes, and the emergent administrative framework designed to balance competing interests. The establishment of the Ranger mine was enabled by the recommendations of the Ranger Uranium Environmental Impact Inquiry (the 'Fox Inquiry') which also considered—and recommended favourably—the establishment of Kakadu National Park, and the recognition of land rights for Indigenous people (Fox, Kelleher and Kerr 1977). Whilst noting the opposition of the Mirrar Gunjeihmi people to the establishment of the Ranger mine, the Fox Inquiry sought a compromise between competing land interests in the region.

The recommendations of the Inquiry established arrangements to ameliorate the impacts of mining on local Indigenous people including the provision of economic benefits through the payment of royalties under the ALRA, and the implementation of complex land tenure arrangements. Notably the mining town of Jabiru that was established to service the mine became a restricted access area in order to minimise interaction between local Indigenous people and the mining community. The arrangements established by the Fox Inquiry are sometimes referred to as a 'social contract'.

The administrative arrangements established by this 'social contract' remain vexed 30 years after the establishment of the Ranger mine. Historically Ranger has had low rates of local Indigenous employment. Despite the payment of royalties under the terms of the Ranger Agreement the Indigenous share of the mine economy is minimal. This is a common pattern at all three mines considered. The impacts on Indigenous people of large-scale development follow the pattern set by Ranger (Taylor and Scambary 2005).

The YLUA is an Indigenous Land Use Agreement between Hamersley Iron (now known as Pilbara Iron) and the Yinhawangka, Banyjima, and Nyiyaparli[1] people of the Central Pilbara. The agreement is not registered with the National Native Title Tribunal. The agreement has a regional focus and concerns an area of approximately 26,000 square kilometres (Senior 2000), encompassing much of the traditional land interests of the three language groups, and a substantial area of Pilbara Iron's mining and exploration tenements in the region. Notably the Yandicoogina deposit is also the subject of mining tenements held by rival iron ore miner BHP Billiton. The Yinhawangka, Banyjima and Nyiyaparli people are parties to a separate agreement with BHP Billiton concerning the BHP Billiton-owned Yandicoogina Iron Ore Mine.[2]

---

[1] The orthography used here accords with that recommended by the Wangka Maya Pilbara Aboriginal Language Centre. However, numerous spellings of these language group names are in use in the region.
[2] The Mining Area C Agreement between Yinhawangka, Banyjima, Nyiyaparli people and BHP Billiton establishes the IBN Corporation with the same membership as Gumala.

The GCA is a Future Act Agreement pursuant to s.29 of the *Native Title Act 1993* (Cwlth) (NTA), and is between the State of Queensland, Century Zinc Limited (CZL) and the Waanyi, Mingginda, Gkuthaarn and Kukatj people of the southern Gulf of Carpentaria. The GCA predates the 'Wik' amendments to the NTA of 1997, also known as the 'ten-point plan amendments'.[3]

The central features of the YLUA and the GCA, like the Ranger Agreement, are preferential concessions relating to provision of employment and training; Indigenous business development; heritage protection and financial recompense for mining; and heavy emphasis on integrating Indigenous people into the mine economy.

However, the dollar amount of the YLUA and GCA, approximately $60 million over the anticipated 20-year life of both mines, is significantly less than royalty payments made by Ranger mine operator Energy Resources of Australia (ERA) through a complex set of arrangements to the Aboriginal Benefits Account (see below) (O'Faircheallaigh 2003a).[4]

Total royalties paid by ERA and distributed to the Aboriginal Benefits Account since the commencement of mining in 1980 are $207.7 million (see also ACIL Economics and Policy Ltd 1993: 17; ACIL Economics and Policy Pty Ltd 1997: 3; ERA 2006). ERA erroneously states that the company has paid this amount to Indigenous interests (ERA 2006); in fact they are paid to the state, which then distributes mining royalty equivalent amounts to Indigenous interests and the Northern Territory Government. Traditional owner groups only receive 30 per cent of these payments via royalty associations that have been incorporated to receive such funds. The Gagadju Association was the nominated organisation to receive such payments from 1979 to 1995. From 1996 to the present the Gunjeihmi Aboriginal Corporation has received all royalty payments from Ranger.

There has been much conjecture over the distribution and expenditure of royalty payments throughout the history of the Ranger project. It is commonly assumed that traditional owners personally receive large sums of money. Over time this assumption has resulted in the non-provision of government services on an equitable needs basis to Indigenous people in the Kakadu region (Kakadu Region

---

[3] The 'ten point plan' amendments, were in direct response to the High Court's decision in The Wik Peoples v The State of Queensland & Ors; The Thayorre People v The State of Queensland & Ors [1996] HCA 40 ('Wik decision'), which found that native title could coexist with pastoral leases. In the event of any conflict the High Court found that the rights of pastoralists would prevail. The intention of the amendments was to seek a compromise for conflicting interests, with Prime Minister Howard claiming that 'The fact is that the Wik decision pushed the pendulum too far in the Aboriginal direction. The 10 point plan will return the pendulum to the centre' (Amended Wik 10 Point Plan, Media Release, 8 May 1997). The amendments had significant beneficial impacts for land access for the mining industry, and were undoubtedly influenced by the Australian Mining Industry Council's sustained campaign for blanket extinguishment of native title rights and interests.

[4] For a description of the complex of royalty distributions arrangements of the Ranger Agreement see Altman 1983: 56–61.

Social Impact Study 1997a; O'Faircheallaigh 2004a), and allegations of profligate expenditure by individuals and organisations alike. Such views are a historical legacy of the Queensland Mines Ltd (QML) Agreement, which related to the nearby Nabarlek uranium deposit. Negotiated after the Ranger Agreement, the QML Agreement saw the distribution of cash payments to members of three associations prior to any distribution of Ranger money in 1979.

The community benefit package of the YLUA was envisaged to include approximately $60 million in cash payment to the Gumala Aboriginal Corporation, representing the 430 members of the Banyjima, Nyiyaparli and Yinhawangka peoples, over the anticipated 20 year life of the Yandicoogina mine. However, increased production in response to world demand for iron ore make it likely that the mine will have a 10–15 year lifespan, and Gumala will be paid significantly more than the anticipated $60 million.

In addition a range of training, employment, heritage protection and business enterprise development initiatives are contained in the agreement, and there are provisions for the staged return of Rocklea station which is owned by the mining company. 'Community' benefits from the agreement are primarily administered via the Gumala Aboriginal Corporation in the form of trust funds known as the General Trust and the Elderly Infirm Trust, the latter ceasing after the first five year period of the agreement. Gumala has developed a business enterprise unit known as Gumala Enterprises Pty Ltd (GEPL), and has entered into joint business ventures in transport, equipment hire and camp management services in the Pilbara region. The trusts are designed to provide assistance to the membership through the delivery of programs such as investments, culture, law, community development, business development and education. The capacity of the trusts to make financial payments to individual members is restricted by the charitable nature of the trusts. In the first five years of operation the trusts received approximately $15.3 million from Hamersley Iron. The lack of direct access to these funds for Indigenous parties to the agreement due to trust arrangements has led many Gumala members to perceive that they have no control over the compensatory benefits derived from the YLUA. Such dissatisfaction is encapsulated in the comment from Banyjima man BF, at the time of fieldwork, when he stated that 'We have the richest trusts, but the poorest people'. As such the existence of substantial trust funds and the poor and declining socioeconomic status of Indigenous people who might lay claim to them is considered an indictment by many of the capacity for agreements such as the YLUA to achieve any substantial economic development or sustainable outcomes for Indigenous people.

The GCA is a complex document that commits the five native title groups, CZL, and the Queensland Government to a relationship that is designed primarily to facilitate the mining and transportation of ore from the Century mine. In addition the objectives of the GCA for Indigenous people include the reduction of welfare dependence, and the promotion of economic self-sufficiency, better health and education standards, access to country, and community and cultural development. Undoubtedly, such aims flow from Indigenous people's symbolic approach to negotiations, and their desire to achieve appropriate recompense for past injustices, including dispossession of traditional lands and subsequent enduring poverty (Blowes and Trigger 1998: 109). Existing Indigenous disadvantage in the region poses serious challenges for realising these goals (Martin 1998a: 4). Failure of the GCA to attain any substantial improvement in the relative disadvantage of Indigenous people, particularly the Waanyi language group, is the subject of intense efforts of the Carpentaria Land Council Aboriginal Corporation to seek amendments to the agreement (Flucker 2003a, 2003b). Such attempts include the conduct of a review of the GCA by the Waanyi Nation Aboriginal Corporation, which followed a nine-day occupation of the mine canteen in 2002 by approximately 200 Waanyi people (see also Martin, Chapter 5; Trebeck, Chapter 6).[5] Like the YLUA the GCA contains provisions for the incremental return of pastoral land holdings owned by CZL.[6]

Arguably the most successful element of the GCA is the average 15–20 per cent employment of local Indigenous people at the Century mine, an employment ratio that eclipses the national average of 4.6 per cent Indigenous employment in the mining industry (Barker and Brereton 2004). In the life of the mine approximately 550 people from the Gulf of Carpentaria have been employed, and between 100 and 120 Indigenous people at any one time between 2001 to the present (Barker and Brereton 2004). Predominantly, Indigenous people are employed in the mine pit as truck drivers and operators, but significant numbers

---

[5] In 2002, after a meeting held at Bidanggu outstation to discuss a review of the GCA, approximately 150 Waanyi men, women and children drove to the Century mine site, announced their presence at the site office, and then occupied the mine canteen. The Queensland Government ordered the mobilisation of the police special squad to the mine site, but was thwarted in its initial efforts to dislodge the protestors by declarations in the media by Waanyi spokespeople that the protestors were unarmed and mainly elderly people and children. Intense negotiations began between the general manager of the mine and the protestors. The sit-in lasted for nine days and severely disrupted the meal routines for the approximately 400 strong fly-in-fly-out workforce. In addition, the unprecedented move to occupy part of the mine site sent shock waves through the business community and threatened to halt production at the world's largest zinc mine ('Aborigine protest threatens zinc mine', K. Meade, *The Australian*, 19 November 2002). The sit-in exposed the mine to serious financial risk which could have been critical for the continuation of the operation given that Pasminco (the operating company at the time) was experiencing financial difficulties at the time. The sit-in arose through dissatisfaction with the perceived limited scope and lack of independence of 'The Five Year Review of the GCA' undertaken by the Gulf Aboriginal Development Company Ltd, Pasminco and the State of Queensland.
[6] Turn Off Lagoon has been returned to Waanyi People, and in 2007 CZL retains a 49 per cent stake in Riversleigh and Lawn Hill Stations. Lawn Hill Station is a commercially lucrative pastoral enterprise carrying 50 000 head of cattle. Both Lawn Hill and Riversleigh stations are managed by the Lawn Hill and Riversleigh Pastoral Holding Company, which currently sublets the properties to non-Indigenous commercial pastoral enterprises and also conducts a pastoral training program in association with these two stations. In addition the Gangalidda people, who are not parties to the GCA, have received title to Pendine and Konka Stations.

are also working in mine administration and service areas associated with the mining camp. Reasons for such high Indigenous employment overall include the operation of Community Liaison Offices in the communities of Doomadgee, Normanton, and Mornington Island, funding by the State of Queensland for mine related TAFE training, and the proactive employment strategies of the major contractor, the Roche Eltin Joint Venture, which operates the mine pit.

The three agreements considered here arise from different legislative and administrative regimes. Differences between the agreements also relate to the local circumstances in which each of the agreements were negotiated. The YLUA and the GCA emphasise the economic engagement of Indigenous people, rather than the payment of royalties as with the Ranger Agreement. However cash payments are made in both the YLUA and the GCA, although they utilise the 'real economy' discourse and emphasise participation in mainstream economic activity.

Across all three agreements significant numbers of intended agreement beneficiaries are unable to participate in programs of employment, training or business development due to their status in relation to development-defined socioeconomic indices. Many Indigenous people who have land interests affected by major mining developments are precluded from participating in the mine economy by chronic health issues, limited education, a criminal record, substance abuse issues or old age (Taylor and Scambary 2005). The status of Indigenous Australians against standard social indicator areas including health, housing, education, and labour force participation are indicative of levels of poverty that deserve the moral outrage reserved for the Federal Government response to the incidence of child abuse in Indigenous communities resulting in the Northern Territory emergency intervention.

However, such moral outrage denies the agency and productive capacity of many Indigenous people that arises from an extensive range of skills and knowledge that lie outside the mainstream economy, and that inform Indigenous responses to poverty. Whilst the under-resourcing of services to Indigenous people is one reason for poverty that prevents participation in the mainstream economy, its day-to-day alleviation is often sought through the use of natural resources and the accompanying corpus of knowledge. Scarce government assistance has been inadequate for decades, entrenching Indigenous disadvantage. In many locations this places extreme pressure on the livelihoods of Indigenous people through institutional exclusion and excessive coercion to participate in the 'mainstream'. The denial of access to land and infrastructure essential to the conduct of customary activities and beliefs has added further constraints on Indigenous livelihoods.

Almost universally Indigenous people seek to shape their economic engagement by utilising their skills as Indigenous people, rather than highlighting the capacity and skills deficit identified by standard social index assessments. This raises the possibilities for alternative forms of engagement by reference to diverse Indigenous aspirations for the future. In the context of mining agreements such aspirations are characterised by a desire for agreements that engender more innovative economic relationships, in both mainstream economic opportunities and in enhancing customary sector economic activity (Altman 2005a).

Modern mining agreements arise from legislative frameworks such as the ALRA and NTA that privilege the continuation of Indigenous traditions in the recognition of rights to land, and provide mechanisms to negotiate agreements with resource developers. However, resulting mining agreements de-emphasise the cultural prerogatives of Indigenous people in favour of mainstream economic development initiatives, predominantly within the mine economy. The capacity and desire of Indigenous people to engage in mine employment and training is influenced by diverse life histories resulting in considerable diversity of residence, access to their traditional lands, education standards, health standards, and customary knowledge and experience.

Indigenous policy and mining agreements

In the 1990s Indigenous policy in Australia changed significantly as the tenets of economic liberalism were adopted, increasing the role of the private sector in Indigenous affairs both in terms of philanthropy and in ways consistent with 'practical reconciliation'. This process has continued in the new millennium with the abolition of representative structures such as the Aboriginal and Torres Strait Islander Commission (ATSIC), the introduction of 'mutual obligation frameworks', and most recently with the Federal Government's intervention into the administration of Indigenous welfare, land, and moral life of Indigenous people in the Northern Territory. Diminishing state capacity and growing demands for state services (Quiggin 2005: 22) have been felt acutely in remote and regional parts of Australia—the areas where mines exist and significant numbers of Indigenous people live. In this context mining agreements with Indigenous people vest considerable 'state-like' powers in the industry in relation to defining social policy and delivery of services in select remote and regional areas of Australia. This sits uncomfortably with the industry as corporations resist the pressure to fulfil the role of service delivery (Mining Minerals and Sustainable Development 2002) which is traditionally the domain of the state—creating uncertainty for Indigenous people residing in mine hinterlands.

In 1991 the Report of the Royal Commission into Aboriginal Deaths in Custody recommended that, in light of the extent of Indigenous disadvantage identified in the course of the enquiry, 'Reconciliation of the Aboriginal and non-Aboriginal communities must be an essential commitment on all sides if change is to be genuine and long term' (Commonwealth of Australia 1991). The report urged bilateral support for its recommendations and in the same year the Council for

Aboriginal Reconciliation was established as a statutory authority under the *Council for Aboriginal Reconciliation Act 1991*. The legislation also set the terms for a process to be conducted over a ten-year timeframe to advance formal reconciliation between Indigenous and non-Indigenous Australians. The critical endpoint for this formal process was set to be the Centenary of Federation in 2001.

Initial articulations of the policy of reconciliation were focused on a rights-based approach and accompanied by events such as the High Court's judgment in Mabo No. 2, and the subsequent passage of the NTA which established a national framework for the recognition of pre-existing Indigenous rights in land. Previous Indigenous policies also came under scrutiny—such as that of forcibly removing Aboriginal children from their families in the context of the Stolen Generations Inquiry.[7]

A change of government in 1996 ushered in a different approach to reconciliation that focused on the attainment of 'statistical equality' under the rubric of 'practical reconciliation'. The Howard government claimed that the symbolic rights-based approach of the previous administration had been unsuccessful. Practical reconciliation seeks to address Indigenous disadvantage in relation to tangible indicator areas such as housing, health, education and employment (Altman and Hunter 2003), whilst downplaying the 'rights' or symbolic reconciliation agenda of the previous administration.

The policy shift to practical reconciliation is evidenced by the amendments to the NTA in 1998 which significantly reduced the extent of rights recognised under the legislation. These amendments were designed to create certainty of tenure for pastoral and mining interests, in light of competing claims of prior Indigenous ownership.

Following the abolition of ATSIC in 2005, 'mainstreaming' became the dominant term for the change in direction of Indigenous policy due to its emphasis on the delivery of services via already established government departments and state mechanisms, and the de-emphasis of existing Indigenous service delivery and representative organisations ('the Indigenous sector'). Features of mainstreaming include the attempted coordination of service delivery across State and Federal agencies, and an emphasis on shared responsibility agreements at the local level based on principles of mutual obligation. A premise of mainstreaming is the notion that 'passive welfare' has had a devastating impact on Indigenous Australians (Rowse 2006: 169). The main proponent of this view is Indigenous leader Noel Pearson, who asserts a four-point plan for the development of a 'real economy' on Cape York Peninsula in Queensland. Pearson's plan entails access

---

[7] In 1995 the Human Rights and Equal Opportunity Commission established the National Inquiry into the Separation of Aboriginal and Torres Strait Islander Children from Their Families, also known as the Stolen Generations Inquiry, which conducted hearings nationally throughout 1995–96.

to traditional subsistence resources, adaptation of welfare programs into reciprocity programs, the development of community economies, and engaging in the real economy (Pearson 2000: 83). The last aspect of Pearson's plan—engagement in the real economy—has been assimilated into Indigenous policy frameworks of both the state and the private sector as equating with middle Australia objectives of mainstream economic engagement. Principle indicators of such engagement are culturally informed and include gainful employment and private ownership of property.

Accompanying this policy shift, or perhaps informing it, there has been a disciplinary shift in policy development away from the humanities and in particular, anthropology, towards economics (Altman and Rowse 2005: 159). Altman and Rowse question whether the variant objectives of Indigenous policy 'achieve equality of socioeconomic status or [...] facilitate choice and self determination' (Altman and Rowse 2005: 159). They indicate that the former is the focus of economically-informed social policy, which downplays 'difference' in favour of equality, whilst traditionally the latter has been based on the advice of anthropology and its emphasis on 'cultural difference'. In this sense culture is something that 'aggregates people and processes, rather than integrates them' (Cohen 1993: 195–6).

The emerging tension in Indigenous policy is central to mining agreements—in particular, how the influence of economic liberalism on practical reconciliation excludes (or at best de-emphasises) the cultural imperatives of Indigenous economic agency. As Altman and Rowse note 'This approach ignores a point made by anthropology: that to change peoples' forms of economic activity is to transform them culturally' (Altman and Rowse 2005: 176).

Mining agreements are one way in which mainstream economic participation of Indigenous people is pursued. While modern agreements promote an ethos of economic participation, they are also subject to considerable variation in the involvement of the state. For example, the Queensland Government is a party to the GCA, and the Federal Government is a party to the Ranger Agreement, whilst the Western Australian Government is not a party to the YLUA. In all three cases the state's primary concerns are to ensure the unimpeded development of mineral resources, and minimise liabilities arising from the impairment of native title (O'Faircheallaigh 2006: 9) and, in the case of the Ranger mine, the ALRA. Limited programs such as the Working in Partnership program are funded by the Federal Government to promote greater participation of Indigenous people in the mining industry (Department of Industry Tourism and Resources 2006). More recently, Closing the Gaps initiatives promote 'place based training initiatives' with the resource sector as part of broader strategies to increase economic participation and to address the divergent life expectancies beween Indigenous Australians and the wider population.

Tension exists between the mainstreaming approach to Indigenous affairs and the substantial Indigenous sector that acts as an interlocutor with the state in the delivery of services (Rowse 2006). Organisations that were established under the ALRA and NTA are critical to the negotiation of agreements. NTRBs are federally funded to represent the interests of Indigenous people within a geographic area under the terms of the NTA. In the Northern Territory, land councils established under the ALRA have assumed responsibility for representation of the native title interests of their constituents within their geographic boundaries, and are also recognised as NTRBs. With the NTA 1998 amendments, NTRBs have experienced a substantially increased workload due to increased complexity in the operation of the Act, and the introduction of strict time frames associated particularly with negotiation processes. Other agencies integral to the carriage of processes under the NTA, including the Federal Court of Australia and the National Native Title Tribunal, have received substantial funding increases to address this increased workload. However, NTRBs have experienced an overall decline in funding, and increased intervention by the Federal Government in the discretionary use of funding. O'Faircheallaigh notes that this has reduced the capacity of these organisations to represent the interests of their clients adequately (O'Faircheallaigh 2006: 11–2). Increasingly, the mining industry and other third party developers are funding NTRBs, and Indigenous people directly to fast-track processes associated with the NTA in order to reach timely development outcomes (Minerals Council of Australia (MCA) 2006). Although such direct funding is aimed at pragmatic outcomes, it raises the serious prospect of a conflict of interest in adversarial negotiations over land use (Morgan, Kwaymullina and Kwaymullina 2006).

A submission to the Commonwealth Government by the peak mining industry organisation, the MCA, notes that 60 per cent of mining operations in Australia are adjacent to Indigenous communities (MCA 2006: 25).[8] The same submission notes that NTRBs 'provide a critical platform for industry to negotiate mutually beneficial outcomes' and recognises that 'NTRBs have been chronically under-resourced in fulfilling their legislative functions in representing Indigenous interests' (MCA 2006: 30). Such a shortfall in resourcing, 'has delayed the negotiation of mutually beneficial agreements and forced mineral companies to meet the resourcing gap' (MCA 2006: 30).

As the mining industry seeks to promote the development of 'sustainable regional communities' beyond the life of the mine (MCA 2006: 23), and via the negotiation of agreements with Indigenous people, the inadequacy of state services in the provision of community infrastructure and social services is increasingly hampering such efforts (MCA 2006: 25; see also Holcombe, Chapter 7). The industry has criticised the government for the increased onus upon it to provide

---

[8] 'Communities' in this sense refers to the physical locations where Indigenous people reside.

such services in the absence of social service provisioning (Mining Minerals and Sustainable Development 2002). A key conclusion is that mining agreements on their own are largely incapable of effecting significant mainstream economic outcomes for the majority of Indigenous people who are parties to them.

O'Faircheallaigh's recent study of approximately 45 mining agreements in Australia suggests that the limited success of such agreements flows from the NTA's weakness as a statutory regime for negotiation (O'Faircheallaigh 2000, 2003a, 2004a, 2006; among other papers that make up the study; see also Altman, Chapter 2; Martin, Chapter 5).

Development, aspirations, and livelihoods

The development ethos that informs the current policy direction in Indigenous affairs, and is a keystone to the formal engagement between the mining industry and Indigenous people in the context of agreements, defines Indigenous people as underdeveloped. Esteva (2005: 7) signals this corollary to post-war development discourse and grounds 'the burden of connotations that it carries' in the language of evolution, growth and maturation. In his historical account of the emergence of development, Esteva emphasises the hegemonic nature of a capitalist project to alleviate perceived poverty and underdevelopment in a colonising and homogenising manner.[9] Esteva (2005: 18) asserts that the social construction of development is integral to an autonomous economic sphere and the generation of scarcity:

> Establishing economic value requires the disvaluing of all other forms of social existence. Disvalue transmogrifies skills into lack, commons into resources, men and women into commodified labour, tradition into burden, wisdom into ignorance, autonomy into dependency. It transmogrifies people's autonomous activities embodying wants, skills, hopes and interactions with one another, and with the environment, into needs whose satisfaction requires the mediation of the market.

Esteva's rejoinder to the coercive dependencies that he identifies as being produced by development and the market economy, is to draw attention to the strategies of the 'common man' at the margins of economic hegemony, to re-embed economic practice in culture, and develop a 'new commons'. Esteva (2005: 21) envisages a cultural revival of sorts, and a reclamation of the definition of needs in the name of reducing scarcity. Culturally embedded education and healthcare, he asserts, removes the need for absent teachers and schools, doctors and hospitals and reaffirms the multiple strategies for survival entailed in Indigenous cultural knowledge and relationships to the environment (Esteva 2005: 20–1). Esteva's work serves as a useful reminder of how alternative modes

---

[9] There is a broad literature criticising 'development' (see for example Crush 1997; Escobar 1995; Hobart 1993; Mehmet 1995; Nederveen Pieterse 1994).

of economic interaction might privilege the skills and capacities derived from Indigenous knowledge systems, in lieu of the skills and capacities conventionally valued by Western industrial measures.

Like Esteva's new commons, critical analysis of the development paradigm has generated a post-development discourse that beckons consideration of non-market economic relations and customary activities as legitimate forms of economically productive action. Gibson and Graham's notion of a 'diverse economy' is premised:

> ...on unhinging notions of development from the European experience of industrial growth and capitalist expansion; decentering conceptions of economy and deessentialising economic logics as the motor of history; loosening the discursive grip of unilinear trajectories on narratives of change; and undermining the hierarchical valuations of cultures, practices and economic sites (Gibson-Graham 2005: 5).

The Gibson and Graham study of the municipality of Jagna in the Philippines identifies a diverse economy consisting of 'a thin veneer of capitalist economic activity underlain by a thick mesh of traditional practices and relationships' (Gibson-Graham 2005: 16) that ground what they term the 'community economy'. They explain this community economy as:

> Those economic practices that sustain lives and maintain *wellbeing directly* (without resort to the circuitous mechanisms of capitalist industrialisation and income trickle down) that *distribute surplus* to the material and cultural maintenance of community and that actively make a *commons* (Gibson-Graham 2005: 16).

Reference to this approach is not to suggest that a return to the primordial past is desired by Indigenous people, but rather that the alterity of Indigenous culturally grounded economic activity is maintained despite the colonial experience. From research conducted over a 25 year period with Kunwinjku people of Western Arnhem Land, Altman has developed a model for the analysis of the interdependencies of the market, the state and the customary components of the economy (Altman 2005a: 36). Altman's hybrid economy recognises the 'intercultural' context of the economy in remote areas where the products of customary activities supplement resources from other sectors. Hunting, gathering and fishing often significantly supplement household and community production and consumption (Altman 1987; Bomford and Caughley 1996; Griffiths 2000), and are supported indirectly by the state, for example in the form of Community Development Employment Projects payments. The production and sale of Indigenous art is informed by cultural knowledge, facilitated by government funded art centres, and driven by profits from a lively international art market (Altman 2005a: 38). Other examples of hybridity include the commercial use of wildlife, cultural tourism, and bio-diversity management (Altman 2005a). Underlying the growing importance of this last factor is increasing global concern for the state of the environment, particularly in terms of climate change and dwindling water resources. The majority of Indigenous Australians reside in urban and metropolitan areas. However approximately 26 per cent of Indigenous Australians, or 120 000 individuals reside 'on what is increasingly referred to as the Indigenous estate, an area that covers about 20 per cent of the Australian continent or about 1.5 million square kilometres mainly made up of environmentally intact desert and tropical savanna' (Altman, Buchanan and Larsen 2007). Increasingly, Indigenous people in these regions are engaging in programs of biodiversity management that utilise Indigenous knowledge systems in the control of exotic weeds and feral animals. Traditional fire management practices particularly in the tropical savannas are being adapted to pastoral management, biodiversity protection, and innovatively in privately negotiated carbon abatement programs (Northern Land Council 2006). Government bodies, such as the Australian Quarantine and Inspection Services and the Australian Customs Service, are forming partnerships with Indigenous people living in remote areas and employing them to undertake important activities including border control and disease management. Such activities are formalising the hybrid economy model espoused by Altman, through increased public sector funding for bio-diversity projects.

Within the current policy debate in Australia that is increasingly asserting the failure of self-determination approaches over the last 30 years, economic liberalism and the pursuit of practical reconciliation has found support for greater market integration from influential Indigenous spokespeople such as Noel Pearson and Warren Mundine. Pearson's 'real economy' model highlights a disjuncture between post-colonial Indigenous cultural dispositions and Indigenous society's capacity to attain development outcomes. Central to Pearson's argument is the concept of 'welfare poison', which he maintains has undermined traditional society and authority and instituted a destructive dependence on the state (Pearson 2000a, 2000b).[10] Pearson's four point plan for the establishment of the 'real economy' shares a number of tenets with both Gibson-Graham's diverse economy, and Altman's hybrid economy (Altman 2007a; Gibson-Graham 2005). But, as Altman notes, Pearson's emphasis upon engagement with the market economy has gained prominence and provides 'moral authority' to the 'pro-growth' discourse of Indigenous development. The intervention of the Commonwealth Government in the Northern Territory through the acquisition of communally owned land, infrastructure, and capital items that are purchased with government money and the introduction of income management schemes for Indigenous welfare recipients has extended the moral authority of the state

---

[10] 'Misguided policies a toxic cocktail', N. Pearson, *The Australian*, 24 October 2000.

into unprecedented involvement in the daily lives and affairs of Indigenous people. Such an outcome is paradoxically contrary to Pearson's vision of reducing the institutional involvement of government in the lives of Indigenous people.

The invisibility of the customary economy masks the value that is derived from the exploitation of land based resources by Indigenous people residing on their traditional estate—and exercising agency in determining their own future. During the 1970s many Indigenous people moved away from government and mission settlements to their traditional lands. The 'homeland movement' was primarily a north Australian phenomenon and was enabled to some extent by policy and legislative developments. Altman (1987) notes that decentralisation assisted in the revitalisation and continued practice of hunter-gatherer technologies and practice. Increasing mineral prospecting, particularly in Arnhem Land in the Northern Territory, and the desire to protect sacred sites was also a motivating factor in decentralisation (Gray 1977).

While the approaches of Gibson-Graham, Pearson and Altman present different understandings of non-market economic activity, overall their work can be characterised as taking a livelihood approach to economic development (de Haan and Zoomers 2005). The term 'livelihoods' refers to the diverse activities in which Indigenous people engage in order to sustain themselves. Livelihoods include tangible economic activities associated with the cash economy including paid employment, welfare and commercial enterprise; and resources from the customary sector derived from activities such as hunting, fishing and gathering. Livelihoods are reliant on networks of relatedness of people to kin and country and entail a complex of obligations defined by a corpus of Indigenous law and custom. In this sense livelihoods encompasses intangible aspects of social life that are reliant not only on physical resources, but also on symbolic resources associated with relatedness to and knowledge of country. These resources are drawn upon constantly in the mediation of the authority of Indigenous individuals within groups, and in the assertion of the distinctiveness of Indigenous identity to the broader world. Livelihood pursuits entail aspects of productive agency aimed at deriving forms of value that are not reducible to an economic analysis. That is, the effort expended in accessing, maintaining and utilising symbolic resources yields definitive constructions of personal and group identity.

Livelihoods are described generally as a range of activities associated with the customary sector, including fishing, hunting, gathering, the production of art and craft, the conduct of ritual, and the maintenance of family and kin relations. Livelihood aspirations emerging from fieldwork undertaken in relation to mining agreements are expressed in terms of the resources perceived to arise from such agreements. They include a range of activities premised on access to and management of land and the development of supportive and representative organisations. Contrary to current policy assertions, access to land is a key Indigenous aspiration. Thus any statement about the centrality of land based relationships and responsibilities is a political assertion of a means of redressing scarcity and social dysfunction associated with living in regional urban environments. In the central Pilbara, Indigenous residents desire access to land for the establishment of family-based 'communities', and the access to resources that residence upon one's own country brings. In the Kakadu region the establishment of a number of outstations was facilitated by the Gagudju Association, which emerged as a successful Indigenous organisation in the context of the establishment of Ranger mine and the declaration of Kakadu National Park. Converse to this positive outcome, Mirrar Gunjeihmi people express their opposition to the development of the nearby Jabiluka deposit in terms of loss of land and hence cultural identity (Gunjeihmi Aboriginal Corporation 2001).[11] In the southern Gulf of Carpentaria access to land, for living-areas and rangelands, is also a key aspiration.[12]

Associated with Indigenous aspirations for access to country are aspirations for a multitude of resources to support such access. Vehicles to get there, funds to build houses, to buy generators and to sink bores, represent some of these tangible and associated aspirations. Access to cash resources to purchase equipment is sought from multiple sources including mining agreement trust funds, government grant funding, and in many cases through labour force participation, or business enterprises.[13] Indigenous aspirations identified can be grouped into a number of general areas that emphasise the interdependencies of models such as those outlined above. The maintenance of family and kin structures reinforces relatedness and rights to land and defines membership, exclusivity and authority within the Indigenous polity, and supports political assertions of cultural distinctiveness. Representative Indigenous organisations present a resource in assertions of rights arising from cultural distinctiveness, particularly when made against the state, and the mining industry. Such organisations are integral in claims to land under relevant statutes, negotiations relating to land access, and in the establishment of partnerships in enterprise development that generate resources required for a broad range of livelihoods. Intra-Indigenous politics and conflict can compromise the efficacy of such organisations to achieve outcomes for their constituents, but also highlight the need for innovative governance design in order to accommodate processes for

---

[11] The Ranger mine town of Jabiru and the Jabiluka leases occupy nearly 50 per cent of the Mirrar Gunjeihmi clan estate (Fox, Kelleher and Kerr 1977: 266; Parliament of the Commonwealth of Australia 1999: 77).

[12] In order to secure its mineral interests in the area, the GCA designates that CZL incrementally return significant land holdings to native title holders (Martin 1998: 6), with CZL maintaining a 1 per cent stake to ensure the availability of this land for future exploration.

[13] All three mining agreements considered here have programs that encourage and facilitate enterprise development.

resolution and management of disputes. A key factor that emerges from examination of mining agreements is the impact that different definitions of 'community' associated with mining agreements can have on the stability of agreement-based Indigenous organisations (see below).

Family and kin structures are intrinsic to the range of pursuits associated with the Indigenous customary economy. Customary rules and norms associated with social relationships influence rights to hunt, fish and gather and to utilise land resources. Such rules and norms are reinforced through the myriad symbolic resources associated with a sentient landscape, and, more formally in many areas, through the conduct of ceremonial activity. Such activities generate a range of social values that identify Indigenous people. Notably, significant numbers of Indigenous people engage in mainstream economic activity without apparent detriment to their sense of identity. For example, a number of Century mine employees indicated their aspiration to obtain 'rangelands'[14] upon which to hunt and live and regarded their employment as a strategic path to gaining the necessary resources to realise this goal. Clearly there is significant diversity within and across the field sites analysed that has not been addressed by the current mainstream approach of the state or the mining industry thus far.

The assumption made here is that value is derived by Indigenous people and groups through culturally informed productive action and serves to create and reaffirm cultural identity, 'which is the fundamental expression of their being' (Throsby 2001: 11). At this point it is useful to consider the terms productivity and value, culture, and cultural identity in more detail. Indeed this fundamental expression is the basis for 'a productive life' (or a good life) and is much greater in its scope than suggested by representations of Indigenous agency in mining agreements. As Povinelli notes:

> Aboriginal notions of work, labor, history, and authenticity are assessed and, in many ways, forged by hunter-gatherer discourses and by Western law, but Aborigines' real-life activities and dialogues also critique and challenge the reified categories of 'hunter-gatherer theory' and produce identity not in any way reducible to them (Povinelli 1993: 27).

Whilst interaction with the mining industry represents only one segment of Indigenous lifeworlds, this forum offers potential benefits, in particular resources that can support and augment the customary economy, by establishing its material and, indeed, symbolic worth through the assertion of cultural difference. However, as a corollary, Indigenous agency is also motivated by a desire to minimise the cost that such engagement may present to expressions of cultural identity. Multiple understandings of how value can be derived underpin the

---

[14] The term 'rangelands' is used in the Gulf region to refer to tracts of land that are available for the pursuit of livelihood activities such as hunting and fishing to the exclusion of other activities such as mining and pastoralism.

choices made by Indigenous Australians and determine the types of productive action taken.

The distinction Altman and Rowse (2005) make between approaches to Indigenous policy grounded in economically informed views emphasising equality and sameness, and approaches based upon anthropologically informed views that emphasise diversity and choice, are indicative of the broader disciplinary relationship in which the role of culture is only recognised within economic systems when it can be commodified. As Throsby suggests, the dominant neo-classical paradigm in economics, which constructs economics as being without a cultural context, is not culture-free. Indeed the economy is a system of social organisation (Throsby 2001: 8–9). Economists employing neoclassical modelling to account for culture, do so only within economic terms and as such 'remain remote from an engagement with the wider issues of culture and real-world economic life' (Throsby 2001: 9). Throsby argues that questions of value are intrinsic to both economics and culture and that they provide a mechanism for the recognition of 'cultural value'. Throsby's definition of cultural value consists of a range of typical characteristics or components including the aesthetic, the spiritual, the social, the historical, the symbolic, and the authentic (Throsby 2001: 28–9). However, whilst cautioning that economic and cultural value must be kept distinct, and that economics has a limited capacity to recognise cultural value in its entirety, he urges that it is 'in the elaboration of notions of value, and the transformation of value either into economic price or into some assessment of cultural worth, [that] the two fields diverge' (Throsby 2001: 41). Indigenous people make assessments of cultural value in accordance with their own traditions, heritage, and institutions. Assessments of cultural and economic value diverge in the context of mining agreements, and are reflected in the structures of the agreements that define the types of available choices Indigenous people can make about the nature of their engagement. Assessments of cultural and economic value then inform emergent relationships between Indigenous people, the mining industry, and the state.

Holistic notions of culture that encompass all facets of the way people do things inevitably encompass economic practice. Many determinist accounts of culture draw relationships between the cultural imperatives of pre-capitalist societies and economic activity. For Throsby 'cultural capital' captures the value of a 'cultural product' (or cultural productivity *per se*), in both its tangible and intangible forms, while recognising the economic and cultural importance of such a product.

The term culture has a myriad of meanings and implications in the popular and academic lexicon. It is useful however, to consider culture in terms of an aggregation of individuals into groups on the basis of shared 'attitudes, beliefs, mores, customs, values, and practices' (Throsby 2001: 4). It assumes that a group's use of 'signs' and 'symbols' to convey meanings is important to the production of its cultural identity (Cohen 1993), and in the sanctioning of the behaviour of individuals both in relation to the group and also external to it. Difference and diversity within the group is implied by the use of the term 'aggregation', which also serves to distance this definition of culture from populist renderings that blur the distinctiveness of cultural groups by assuming homogeneity within them.

Cultural identity implies that an association of individuals is defined by a set of common characteristics, and that the group is reliant on symbolic transactions, and mutual identification. As with culture, cultural identity depends upon symbolism, derived from everyday life, and productive action—as Povinelli asserts in relation to the Belyuen (Povinelli 1993). Individuals are 'active in the creation of culture rather than passive in receiving it' (Cohen 1993). As Cohen notes the action of individuals in developing culture, has implications for the politicisation of cultural identity. He asserts that cultural identity is a matter of autobiography in that 'when we consult ourselves about who we are, it involves more than a negative reflection of who we are not' (Cohen 1993: 198); it also entails context specific judgments and choices, that reflect mutual understanding of signs and symbols. This kind of activity is designed to assert inward identification with a group, and distinction from other cultural identities.

The invisibility of the customary economy when perceived through mainstream notions of the 'productivity', or productive labour, of Indigenous people in the 'customary sector' (Altman 2001a: 5) limits the value that can be derived both by the mining industry and Indigenous people from their mutual engagement. To explain, Indigenous productivity is steeped in cultural continuity and is an integral mechanism for the production of cultural identity (Povinelli 1993). The value of Indigenous productivity in the customary economy is realised through multiple activities including quantifiable pursuits such as hunting and gathering, and the production of art (Altman 2005a); and in less quantifiable activities such as development and maintenance of outstations, engagement in family or kin relations, conduct of ceremony or by engaging with a sentient landscape in the production, reproduction, and reinterpretation of cultural identity, as Povinelli notes by 'just being there' (Povinelli 1993: 31). The quantifiable activities Altman outlines are not productive in a purely economic sense, rather as Povinelli observes:

> it is a form of production in the fullest cultural and economic sense of this term, generating a range of sociocultural meanings and political-economic problems and rewards. Hunting and gathering grounds Belyuen Aborigines' relationship to the Cox Peninsula and, vis-à-vis other ethnic groups in the region, [and] defines their Aboriginality (Povinelli 1993: 26–7).

Such cultural value can only be truly realised by those who produce it, and those who receive it. However, manifestations of the nature or essence of cultural value are readily identifiable in the chains and modes of interaction between Indigenous people and the mining industry. For example, statements about the lack of desire to work for the mining industry in the Pilbara and which are supported with statements about the damage that mining does to the country, or the preference to work 'for my community instead', clearly demonstrate a set of priorities, the pursuit of which entails assessments of value, or cost. Similarly in the Kakadu region the Mirrar Gunjeihmi people mounted an international campaign against the development of a second uranium mine on the Jabiluka lease adjacent to the Ranger Uranium Mine lease.[15]   Opposition to the establishment of the Jabiluka mine by the Mirrar Gunjeihmi is clearly articulated in terms of the cost that has been incurred by them as a result of the Ranger uranium mine. The anti-Jabiluka campaign in 1998 invoked the authority of Indigenous identity in opposition to the threats and constraints presented by mining to their cultural autonomy. Similarly at the Century mine the success of local Indigenous employment programs demonstrates not only Indigenous access to employment, but also local Indigenous people's desire to work there.

When considering cost and value, it is important to determine what motivates people in making a choice about the terms and nature of their engagement. Throughout this chapter the link between Indigenous productive action and cultural identity is implicit; this relationship is not necessarily quantifiable in economic terms, Nonetheless it is observable in the relationships between stakeholders associated with mining agreements.

Cohen notes that a minimal condition for the politicisation of cultural identity is individuals' realisation that ignorance of their culture undermines their integrity, and that such marginalisation creates power imbalances with respect to the marginalisers (Cohen 1993: 199). He notes that culture is expressed symbolically, and as such has no fixed meaning, and that may make it invisible to others. Both Povinelli and Merlan note that Indigenous culture is represented in Australia in popular discourse through legislative and policy frameworks (Merlan 1998; Povinelli 1993). Politicised cultural identity mitigates against reified notions of Indigenous identity, by drawing on a symbolic repertoire to assert distinctiveness. Politicised cultural identity also strives to protect and maintain the body of symbolic resources required for the continued construction and reinterpretation of culture, and which reified notions of Indigeneity are perceived to threaten. For example, in the Pilbara some individuals perceived full-time work in the mining industry as jeopardising the attainment of Indigenous aspirations for the future by placing barriers between them and symbolic resources central to their identity as individuals and as Indigenous

---

[15] For detail on the Jabiluka campaign by Mirrar Gunjeihmi see Trebeck, Chapter 6).

people. Working a twelve-hour shift means distance from family and country and its symbolic value. Conversely, Pearson's claim of the destructiveness of 'welfare poison' assumes the erosion of culture that such dependency has inured, and views the real economy as a means of re-establishing the role of individual responsibility.

Relationships between Indigenous parties to agreements and the industry are never definable purely in terms of Indigenous people's desire or lack of desire to engage in programs provided under the rubric of 'community benefits'. In struggling to maintain links between the present and the mythic and historical past, in the pursuit of aspirations for the future, many older people suggest that young people should both engage with the mine economy, and fulfil obligations of a cultural nature. The need to garner resources from multiple sources—including wage labour, compensation, business development, and engagement in Indigenous cultural and social life—are seen by many Indigenous people as essential in maintaining and augmenting cultural identity. A parallel can be drawn between Richard Davis' observations in relation to Indigenous pastoralists in the Kimberley when he states that 'commercial pastoralism allows Aborigines the capacity to accrue the social and cultural capital that has historically rested with white pastoralists whilst maintaining a radical alterity to them' (Davis 2005: 58). Such alterity is demonstrated by BL, a Kaiadilt man from Bentinck Island in the Gulf of Carpentaria, who, whilst working at Century mine, also maintains a radical opposition to a cyclone-mooring buoy associated with the Century port facility at Karumba, and constructed on a sacred site within his traditional estate.[16] Reconciliation of Indigenous alterity with participation in the mine workforce is highlighted in the statement made by a Gangalidda worker at the Century mine, when he stated his goal as 'helping my people achieve their white dreams but staying black to do them'.

The incorporation of Indigenous values within a mining context is difficult, but clearly not insurmountable. A major obstacle to such incorporation is the industrial disposition of the mining industry that struggles to accommodate cultural diversity within its corporate framework. Similarly, the incorporation of mainstream economic values in Indigenous lifeworlds is difficult, but not insurmountable—the principle obstacles being institutional exclusion that creates incapacity to engage through the fostering of a skills deficit, and by placing barriers between individuals and their cultural identity that the majority of Indigenous people find unacceptable.

However, Indigenous people continually seek to influence both industry and the state to accept modes of engagement that allow for the augmentation of

---

[16] See 'Mad about the buoy', D. Marr, *The Good Weekend*, 18 August 2001.

Indigenous identity, and hence the derivation of value from cultural, political and economic arenas of Indigenous life.

Adverse relations between the mining industry and Indigenous people arise from fundamentally different interpretations of land and its resources. Many Indigenous people associates with his study characterised 'country' as sentient and meaningful, producing socially embedded management practices that yield both tangible and symbolic resources. This contrasts with a non-Indigenous view of landscape that through the exploitation of its resources, becomes socially embedded, and therefore meaningful. Both views contain judgements about the productive value of land, and in turn the knowledge and capacity required to attain such value. However, the two views are not incontrovertible. A clear example of this is the 1946 Pilbara pastoral workers' strike which saw approximately 800 Indigenous pastoral workers simultaneously walk off 25 pastoral properties. The strikers sustained themselves through the prolonged industrial action by engaging in lucrative independent mining activity. The miners organised themselves collectively into successful corporations that accounted for traditional land affiliations and decision making processes.[17] Wilson characterises the strike and the emergent mining operations as the 'Pilbara Aboriginal social movement', which he states was implicitly a claim to citizenship rights (Wilson 1961: 97), occurring within a broader context of Indigenous struggles for recognition and rights (see Attwood 2003). State policies of protectionism and assimilation have enabled the historical dominance over Indigenous interests—firstly of the pastoral industry, and later the mining industry—a situation that endures in the current era. This subordination of Indigenous interests, as highlighted by the Pilbara pastoral strike, obstructs the visibility of Indigenous agency in deriving value from the land, and suppresses the possibilities of the divergence of notions of cultural and economic value in the context of mining agreements, but also more broadly.

The reluctance of the industry to assume responsibilities of service delivery is an increasing source of tension in the tripartite relationships entailed in mining agreements, and one that suggests an emerging nexus between Indigenous consent to mining and access to non-discretionary citizenship rights. In the context of mining agreements, trust funds and in-kind support from the mining industry are increasingly funding Indigenous health, housing and education programs in order to produce the competencies required for mine employment. This emerging role of the industry is converse to the Indigenous sector which is the target of funding cuts, increased scrutiny from government oversight, and devolution of functions to mainstream government departments. Given the

---

[17] A key figure in the strike movement was Don McLeod, controversial as an activist and for his links with the Australian Communist Party. Some accounts credit McLeod with masterminding the strike movement and subsequent successful mining collectives that grew out of it, though Wilson (1961, 1980) gives a more nuanced account of McLeod's influence.

historic dominance of mining interests over Indigenous interests a key consequence of agreements such as the YLUA and the GCA is the inculcation of Indigenous people residing in mining areas into the narrow agenda of mineral development.

In the context of the three agreements considered, the relationships between the mining industry and regional land councils and NTRBs can be characterised as fraught. Such relationships arise from the historic opposition of the industry to the statutes under which these organisations operate, and the constraints they place on mineral development. Amendments to statutes such as the ALRA and the NTA have favoured the industry in its pursuit of security of tenure for commercial mining operations. Past representations by the mining industry of land councils and NTRBs as recalcitrant, anti-development, and self-interested in debates about the workability of legislative mechanisms were reflected in the Howard Government's mainstreaming approach of seeking to bypass intermediary Indigenous organisations in service delivery (Vanstone 2005). However, ongoing management of the agreements considered here, and the relationships that they engender, suggest a clear role for such organisations, and the development of specific and local expertise to represent Indigenous people in dealings with the mining industry. This is despite the varying roles played by such organisations in the three agreements, and the varying support they currently draw from their Indigenous constituents.

The representative and governance expertise of existing organisations is variable across the three regions, and is influenced by the relatively recent establishment of many NTRBs and land councils (for example, the Pilbara Native Title Service), and diminishing levels of resources available through state funding (O'Faircheallaigh 2006). Also, the adversarial relationships between land councils and NTRBs and the mining industry have led to situations where representative organisations are bypassed or heavily criticised, as in the Century mine negotiations.

An enduring example of organisational dysfunction that has influenced modern mining agreements is that of the Kunwinjku Association and the Nabarlek Traditional Owners' Association emerging from the QML Agreement (1979–95) with the Northern Land Council concerning the Nabarlek uranium mine in western Arnhem Land. Research documenting the poor governance of these associations, and unclear definitions of agreement recipients, indicates that significant sums of money derived from the agreement were wasted (Altman and Smith 1994; Kesteven 1983; O'Faircheallaigh 1988). The most recent of these research publications is the review of the Nabarlek Traditional Owners' Association undertaken by Altman and Smith (1994) on behalf of the Northern Land Council, and subsequently published. Altman and Smith's research notes a number of factors that are still relevant, including: the finite life of mines and, hence, financial flows to Indigenous people; the limited capacity for strategic responsiveness to organisational capacity shortfalls of Indigenous agencies by government departments, the mining industry and Indigenous representative organisations; and legislative ambiguity in the purpose of mining money to be compensatory or benefit sharing payments, hence public or private, in their application. Such factors indicate, as does this chapter, that consideration of poor governance and poor accountability in the context of mining agreements extends beyond the Indigenous sector to include the mining industry and the state. Altman and Smith note that the Northern Land Council was subject to intense and perhaps excessive scrutiny for its role in the QML Agreement compared to other stakeholders, but commend the organisation for sponsoring and allowing publication of the research 'in the interests of learning from the mistakes of the past' (Altman and Smith 1994: 1). Such reflexivity is rare in the assessments of mining agreements by stakeholders. Though this example is dated and singular, it nonetheless is transparent and demonstrates the potential and desire of Indigenous organisations in furthering engagement with external parties—a capacity that has in many cases been obscured by the oppositional stances by the state and the mining industry to the discourse of Indigenous rights that inevitably accompanies the representation of Indigenous interests.

Altman and Smith's (1994) research occurred on the cusp of a new agreement era emerging from the passing of the NTA, and coincided with the new approach of Rio Tinto to work with the new legislative framework (L. Davis 1995), an approach subsequently adopted by the industry.[18] The business case for the new approach is influenced by the maintenance of corporate image through the portrayal of mining companies as good corporate citizens (Trebeck, Chapter 6). Obtaining a social licence to operate is premised on companies working in partnership with communities in the areas in which they operate and the generation of community benefit. The Nabarlek case is considered a worst-case example, and one which if repeated might reflect poorly on mining companies as contributing to Indigenous social dysfunction. The new approach of attaining Indigenous community benefit through mining agreements entailed a reactionary response to the perceived wastage associated with the royalty regime under the ALRA, which is extrapolated from Nabarlek to apply generally to all mining agreements under the ALRA. The post NTA practice of tying compensatory payments to specific purposes defined in mining agreements as 'community benefits packages', mitigates against Nabarlek type situations, but also reduces Indigenous autonomy over funds derived from what are essentially commercial

---

[18] However, this new cooperative approach did not deter the industry from strenuous lobbying for amendments to the NTA in 1998 to provide commercial certainty over the pastoral estate that covers 40 per cent of Australia, resulting in the 'ten point plan' (see footnote 2 above).

negotiations under the NTA.[19] The dominant role of the mining industry in setting the terms of modern mining agreements was enabled by the unexpected High Court decision in Mabo No. 2; subsequent uncertainty under the new NTA legislative regime; its national focus; and the absence of Indigenous organisations in many parts of Australia that could assume the mantle of being representative bodies under the NTA.

Community

Expertise to identify people whose land interests and lives are impacted by mining development is reliant on knowledge of local tenure systems, political and social allegiances. Land councils and NTRBs, despite any current shortcomings, are ideally institutionally situated to fulfil this role and subsequent representation of Indigenous interests in negotiations with resource developers.

Associated with all three agreements considered here are incongruous definitions of the relevant community that arise from initial assessment of land interests impacted or affected by the mine. The renegotiation and realignment of the community of benefit by Indigenous people themselves at all three locations has emerged over time. At Ranger mine the Gagudju Association was created to represent the interests of all Indigenous people affected by the mine. Membership of this association was expansive. However, over the life of the agreement relationships amongst the membership have forced a renegotiation of the community of impact and the emergence of a new organisation—the Gunjeihmi Aboriginal Corporation—with a discrete membership comprising of the land owning group subject of the Ranger and Jabiluka lease areas. In the Pilbara, the Gumala Aboriginal Corporation represents an alliance between three language groups. This alliance is threatened by the negotiation of discrete mining agreements outside the Gumala coalition, but within the country of, and on behalf of the membership of the three language groups; and the increasing assertions of language group autonomy over substantial funds held in trust by Gumala. In the southern Gulf of Carpentaria the Indigenous parties to the GCA are members of four language groups. However, the Wellesley Islands Sea claim concluded that Gangalidda people, who are not formally a party to the GCA, had succeeded to the country of Mingindda people, who are a party to the GCA. It follows then that legal definitions of the affected community have been altered significantly.[20]

---

[19] A parallel can be drawn with the restrictions placed on agreement derived expenditure in the 1982 agreement between Pancontinental and the Northern Land Council concerning the Jabiluka prospect (Altman 1983).
[20] A corollary to the inflexibility of the respective agreements to accommodate change membership dynamics is the changed corporate identity of mine operators as a result of corporate takeover at all three locations.

The assumption that initial definitions of the community will encompass the unity of the group for the life of an agreement is inevitably challenged by influences internal to the group, and also by extraneous pressures, as noted above. This is particularly so where community definitions were initially expansive as at Ranger, or entail coalitions as in the Pilbara. Indigenous regional land interests and the intra Indigenous relationships and politics they entail are dynamic and context specific. The process of defining the region of impact of mining varies from region to region and mine to mine and is affected by a number of considerations including: the extent of Indigenous knowledge of local land interests, particularly in urban and regional areas; the scale of development as in the Pilbara; or the potential environmental impacts as in the southern Gulf of Carpentaria and in Kakadu National Park. The YLUA and the GCA statically define the community, and despite five year reviews of both agreements, there appears little scope for reflecting local re-assessments of the relevant community. However, re-definition of the community risks the loss of relevance and efficacy of static agreement structures and organisations in the attainment of agreement outcomes. The decline of the Gagudju Association is a clear example of the adverse impact of the assertion of discrete rights and interests. Inadequate consideration of discrete land interests in the GCA was also a factor in the 2002 Century mine sit-in, demonstrating that mines themselves are not enclaves isolated from the lives of Indigenous people. The mining industry's management and understanding of the complex Indigenous politics associated with land interests and access to benefits is limited. The absence of organisations that possess the relevant specialist expertise to represent Indigenous interests risks drawing the industry further into realms outside its core commercial functions, and raises the potential for conflict.

Undoubtedly the lack of definition of discrete land interests of Indigenous groups party to mining agreements, and the incumbent lack of visibility of rights and interests arising from localised land tenure, limits the workability of agreement governance, and ultimately agreement outcomes. This is not to suggest that agreements cannot be reached on a regional basis, but rather that such agreements must account for local land interests that they encompass in order to maintain their relevance and regional legitimacy.

The objectives of mining agreements to attain regional economic development outcomes are also constrained by the financial scale of the multi-year agreements themselves. In 2003 the Gunjeihmi Aboriginal Corporation received approximately $1.17 million of the approximately $7.6 million annual payment of Ranger Uranium Mine royalties paid to the Aboriginal Benefits Account. Individual payments accounted for approximately $500 000; when divided between the approximately 240 royalty recipients, this resulted in an annual payment of approximately $2400 per person (see also Altman 1997: 180). A further $20 000 was allocated for whitegoods, furniture and other household items across the membership. The remaining $1.5 million was allocated to a range of social services for the membership and the region, including aged care facilities, purchase of one community vehicle, and infrastructural and consumable outstation support (including repairs, maintenance and fuel for generators for example). Money was also allocated for investment and administration for the Corporation itself (Gunjeihmi Aboriginal Corporation 2003).

Whilst there are four language groups who are party to the GCA, the objectives of the agreement seek to positively influence the socioeconomic status of Indigenous people living in the region. The regional Indigenous population is estimated to be 6000 (Earth Tech 2005: 20), whilst an estimate of the membership of the language groups is approximately 900 people. A crude calculation of the $60 million value of the mine over a twenty year period gives an annual expenditure of between $500 and $3300 per person per annum. This does not account for at least $30 million of the agreement funds being dedicated by the State of Queensland to the development of mine/regional infrastructure such as the mine access road. Consideration of this would halve these figures.

Similarly the YLUA provides for approximately $60 million for 430 people over the anticipated 20 year life of the Yandicoogina mine, though these funds are tied to stringently controlled trust funds that are not generally accessible to the membership.

Programs of employment and training, business development, heritage and environmental protection target compensatory benefits at tangible outcomes. Indigenous support for these programs is premised on the economic and social advantage that they represent, but also on the accessibility of such programs to the intended agreement beneficiaries. Inaccessibility of such programs diminishes their relevance in the repertoire of available resources, and generates ambivalent responses. At the Century mine, Indigenous employment is viewed as a successful outcome from the agreement, but one which was undermined by the poor workforce representation of local/mine adjacent Waanyi people, and compounded by dysfunctional agreement structures culminating in the confrontational/activist 2002 sit-in at the mine canteen discussed above. In the Pilbara, the perception of there being no clear progression into employment at the conclusion of training programs, poor land access, and lack of access to trust funds generates Indigenous ambivalence. At Ranger, the request by Mirrar Gunjeihmi to ERA not to employ Indigenous people at all reflects their anti-mining stance, but also a desire to avoid the negative social consequences associated with an influx of Indigenous migrants. Mirrar Gunjeihmi opposition to the development of the Jabiluka prospect also reflects an assessment of cost incurred in terms of reduced cultural autonomy and enhanced social dysfunction as a result of their experience of the Ranger mine. The perception that mining agreements bring prosperity to Indigenous people is promulgated by the mining industry and the state to reduce opposition. However, the documented experience of Indigenous people impacted by the Ranger mine is that cost shifting from the state to both the regional recipients of 'areas affected' payments and the mining company ERA has resulted in the region being arguably economically and socially worse off than nearby comparable regions of the Northern Territory (Kakadu Region Social Impact Study 1997a; Taylor 1999). The withdrawal of the state in the delivery of what should be non-discretionary citizenship rights in the Kakadu region should be cautionary, both to the attainment of mainstream economic objectives implied by modern mining agreements, and the mainstream 'closing the gaps' approach of current Indigenous policy.

Conclusion

Many people who participated in this research maintain that they have seen little benefit from mining agreements. This is largely due to their relative socioeconomic status (see for example Taylor and Scambary 2005). This is not to deny the in-kind assistance and programs that mining companies have engaged in at all three regions. Rather it suggests that the anticipated outcomes associated with the agreements, and arising from varied and complex negotiating processes, have not eventuated. In the Pilbara for example, the existence of a substantial trust fund associated with the YLUA is viewed positively, but the current inability of the Gumala membership to readily access funds creates the widespread perception that they have little autonomy within the agreement to determine the shape of their future. Conversely, in the Gulf of Carpentaria, poor corporate governance associated with the GCA and unstable recipient organisations do not assist in the creation of a capital base, and similarly undermine intended agreement outcomes. The history of the Ranger mine and associated Indigenous organisations highlights the loss of autonomy of the Mirrar Gunjeihmi through the dispersal of their authority in the administrative frameworks designed to minimise the impacts of mining. However, through the interplay of local identity politics associated with Ranger mine and later the Jabiluka protest, the Mirrar Gunjeihmi have re-emerged as powerful actors in the region. This has had consequences for other regional interests, especially the neighbouring Bunidj and Murrumburr people, through the dilution of their authority in the organisations and institutions associated with mining in the Kakadu region.

Despite the provision of mainstream economic opportunities, access to land remains a critical issue at all three sites considered. In the Kakadu region the Ranger and Jabiluka leases occupy approximately 50 per cent of the Mirrar Gunjeihmi estate. Whilst the YLUA and the GCA make provision for the return of pastoral land holdings of the respective mining companies, the outcomes and equity of such provisions are unclear to many. Title to a number of leases in the Gulf of Carpentaria has been granted, though the continuation of commercial pastoral operations on Lawn Hill Station is seen by some to preclude Indigenous use of the area. In the Pilbara, the timeframes for the return of leasehold land is unclear. The desire to access land for livelihood and religious pursuits is a central finding of this research and one that suggests the need to broaden the terms of engagement entailed in mining agreements.

Tangible livelihood outcomes are economic, and are considered by many Indigenous people to be reliable in comparison to the risks of dependency associated with obtaining resources through engagement with the market and the state. Symbolic resources are also derived through the conduct of livelihood activities and are central to the maintenance and construction of distinct Indigenous identities. The symbolism of everyday life is drawn upon in the inward assertions of identification with a group of people, and outwardly in the assertion of difference to other cultural identities. Examples are provided from all three field sites of the continued practice of livelihoods associated with the customary sector. Whilst the yield of livelihood practices has not been quantified here, the nature of cultural value that is derived from such activities is manifest in the choices individuals make about their lives and their limited level of engagement with the mine economy. Assessments of costs and benefits are considered in terms of economic gain, but also in terms of personal and group identity. As such, cultural value derived is only truly perceptible to those who produce it. However, this is not to suggest that poor outcomes against agreement objectives are reducible to the choices Indigenous people make about the nature of their engagement with the mine economy. Rather, the choices people make are to a large extent dictated by the opportunities that are available. This chapter has highlighted the structural obstacles to mainstream economic engagement presented by poverty, social and economic exclusion, and structures arising from the agreements themselves that define narrow terms of engagement. Like the pastoral industry before it, the obstacles to cultural autonomy that are presented by the presence of the mining industry also impact on the customary sector. Access to land—for the purpose of establishing residence, or accessing resources, or maintaining links to important sites in the sentient landscape—is a key aspiration across the three field sites, and is integrally linked to the range of tangible and symbolic resources that land access provides.

The possibility for a convergence of economic value and Indigenous cultural value is clearly reflected in the dual aspirations of Indigenous people across the breadth of this study to enhance both their market engagement, and engagement with the customary sector. Across Northern Australia the recognition of cultural value in financial terms is emerging in innovative partnerships that emphasise Indigenous land management practices. Extensive networks of Indigenous ranger groups are already involved in projects associated with the maintenance of biodiversity, disease control, border control, feral animal and weed management, fire management, and green house gas abatement. Such projects recognise and enhance the value of Indigenous knowledge and capacity deriving from relationships to land, and have the potential for developmental outcomes in terms of the generation of economic resources. The benefits to Indigenous people, aside from those arising from fee-for-service arrangements, include the opportunity for the continuation of cultural traditions, the maintenance of heritage, and the maintenance of distinctive identities.

The possibility for the application of 'community benefit' packages associated with mining agreements in areas of land management is noted in this chapter, particularly given the extensive pastoral land holdings of mining companies. This is not to suggest that programs of mainstream engagement aimed at the mine economy should be abandoned, but rather to suggest a possible area in which the application of community benefits can be fruitfully augmented. Further innovation is required in the forms of engagement between the mining industry and Indigenous people to promote Indigenous empowerment and autonomy. Central to this is the recognition of who Indigenous people are, and respect and accommodation for the diverse range of knowledge and skills they possess.

The aspirations of Indigenous people associated with the Ranger mine in the Kakadu region, the Yandicoogina mine in the central Pilbara, and the Century mine in the southern Gulf of Carpentaria are multiple and diverse. Emerging from the broad scope of this research are numerous Indigenous narratives concerning distinctiveness, authenticity, equality, autonomy and responsibility. These narratives reach beyond the local relationships with the mining industry and state entailed in mining agreements, and draw upon the historic experiences of Indigenous people to demand both citizenship rights and symbolic rights. Indigenous struggles to seek redress of social and economic exclusion draw upon normative modes of social transaction and cultural process that rally against reified representations of indigeneity, and suggest ongoing cultural transformations. Such transformations are reflected in strategies and aspirations for the future that seek to innovatively resolve conflicting notions of productivity and value through positive assertions of Indigenous distinctiveness within the broader realm of an Australian national identity.

Through examination of three mining agreements it is clear that Indigenous people residing in mine hinterlands engage and respond to global influences while at the same time engaging with the customary. A clear example is the Century mine workers who drive haulpac trucks in the mine, but still draw upon the tangible and symbolic resources of their country in the construction of identity and the maintenance of tradition. This chapter has drawn on research associated with three mining agreements across Australia to demonstrate how agreement outcomes are constrained by the very limitations that they place on the agency of the Indigenous people they seek to engage. This chapter suggests that successful engagement between the mining industry and local Indigenous people who reside in mine hinterlands is dependent on accommodation of existing Indigenous skills and knowledge. Examples abound from across all three locales of Indigenous people successfully striving to engage in multifaceted ways with the mainstream economy, and the mine economy, whilst not compromising their innate cultural identity. Poor understanding of Indigenous capacity by the state and the mining industry perpetuates dichotomous relationships with Indigenous people. Such relationships, combined with the historic under-funding of services by the state, and the lack of recognition of the citizenship rights of Indigenous people, limits the capacity for economic and social engagement, and compounds Indigenous poverty.

References

Aboriginal and Torres Strait Islander Social Justice Commissioner (ATSISJC) 2006. *Native Title Report, 2005*, Human Rights and Equal Opportunity Commission, Sydney.

—— 2007. *Native Title Report, 2006*, Human Rights and Equal Opportunity Commission, Sydney.

ACIL Economics and Policy Ltd 1993. The Contribution of the Ranger Uranium Mine to the Northern Territory and Australian Economies: The report of a study for Energy Resources of Australia, ACIL Economics and Policy Ltd, Canberra.

—— 1997. Economic Flows from the Ranger Uranium Mine to Aboriginal Communities in the Northern Territory: A report to Energy Resources of Australia Ltd, Acil Economics and Policy Ltd, Canberra.

Agle, B., Mitchell, R. and Sonnenfeld, J. 1999. 'Who matters to CEOs? An investigation of stakeholder attributes and salience, corporate performance, and CEO values', *Academy of Management Journal*, 42 (5): 507–26.

Ah Kit, J. 2004. 'Why do we always plan for the past? Engaging Aboriginal people in regional development', Paper presented at the *Sustainable Economic Growth for Regional Australia (SEGRA) 8ᵗʰ National Conference*, 7 September, Alice Springs.

Altman, J. C. 1983. *Aborigines and Mining Royalties in the Northern Territory*, Australian Institute of Aboriginal Studies, Canberra.

—— 1985a. *Report on the Review of the Aboriginals Benefit Trust Account (and Related Financial Matters) in the Northern Territory Land Rights Legislation*, AGPS, Canberra.

—— 1985b. 'The payment of mining royalties to Aborigines in the Northern Territory: compensation or revenue?', *Anthropological Forum*, 5 (3): 475–88.

—— 1987. *Hunter-Gatherers Today: An Aboriginal Economy in North Australia*, Australian Institute of Aboriginal Studies, Canberra.

—— 1988. *Aborigines, Tourism and Development: The Northern Territory Experience*, North Australia Research Unit, ANU, Darwin.

—— 1997. 'Fighting over mining moneys: The Ranger Uranium Mine and the Gagudju Association', in D. E. Smith and J. Finlayson (eds), *Fighting over Country: Anthropological Perspectives*, CAEPR Research Monograph No. 12, CAEPR, ANU, Canberra.

—— 2001a. 'Economic development of the Indigenous economy and the potential leverage of native title', in B. Keon-Cohen (ed.), *Native Title in the New Millennium*, Native Title Research Unit, AIATSIS, Canberra.

—— 2001b. 'Indigenous communities and business: Three perspectives, 1998–2000, *CAEPR Working Paper No. 9*, CAEPR, ANU, Canberra.

—— 2004. 'Economic development and Indigenous Australia: Contestations over property, institutions and ideology', *The Australian Journal of Agricultural and Resource Economics*, 18 (3): 513–34.

—— 2005a. 'Development options on Aboriginal land: Sustainable Indigenous hybrid economies in the twenty-first century', in L. Taylor, G. Ward, G. Henderson, R. Davis and L. Wallis (eds), *The Power of Knowledge, the Resonance of Tradition*, Aboriginal Studies Press, Canberra.

—— 2005b. 'Economic futures on Aboriginal land in remote and very remote Australia: Hybrid economies and joint ventures', in D. Austin-Broos and G. Macdonald (eds), *Culture, Economy and Governance in Aboriginal Australia*, Sydney University Press, Sydney.

—— 2006. 'In search of an outstations policy for Indigenous Australians', *CAEPR Working Paper No. 34*, CAEPR, ANU, Canberra.

—— 2007a. 'Alleviating poverty in remote Indigenous Australia: The role of the hybrid economy', *Development Studies Bulletin,* 72 (March): 47–51.

—— 2007b. 'Indigenous rights, mining corporations and the Australian state', Paper presented at the United Nations Research Institute for Social Development Workshop *Identity, Power and Rights: The State, International Institutions and Indigenous Peoples*, 26–27 July, Geneva, Switzerland.

—— 2008. 'Indigenous rights, mining corporations and the Australian state', Australian case study for the United Nations Research Institute for Social Development project 'Transnational Governmentality of Resource Extraction', Canberra.

——, Biddle, N. and Hunter, B. H. 2005. 'A historical perspective on Indigenous socioeconomic outcomes in Australia, 1971–2001', *Australian Economic History Review*, 45 (3): 273–95.

——, —— and —— 2008. 'How realistic are the prospects for "closing the gaps" in socioeconomic outcomes for Indigenous Australians?', *CAEPR Discussion Paper No. 287*, CAEPR, ANU, Canberra.

——, Buchanan, G. and Biddle, N. 2006. 'The real 'real' economy in remote Australia', in B. H. Hunter (ed.), *Assessing the Evidence on Indigenous Socioeconomic Outcomes: A Focus on the 2006 NATSISS*, CAEPR Research Monograph No. 26, ANU E Press, Canberra.

——, —— and Larsen, L. 2007. 'The environmental significance of the Indigenous estate: Natural resource management as economic development in remote Australia', *CAEPR Discussion Paper No. 286*, CAEPR, ANU, Canberra.

—— and Dillon, M. C. 1988. 'Aboriginal land rights, land councils and the development of the Northern Territory', in D. Wade-Marshall and P. Loveday (eds), *Northern Australia: Problems and Prospects Volume 1: Contemporary Issues in Development*, North Australia Research Unit, ANU, Canberra.

—— and Hunter, B. H. 2003. 'Evaluating Indigenous socioeconomic outcomes in the reconciliation decade, 1991–2001', *Economic Papers*, 22 (4): 1–15.

—— and Johnson, V. 2000. 'CDEP in town and country Arnhem Land: Bawinanga Aboriginal Corporation', CAEPR *Discussion Paper No. 209*, CAEPR, ANU, Canberra.

—— and Levitus, R. 1999. 'The allocation and management of royalties under the Aboriginal Land Rights (Northern Territory) Act: Options for reform', *CAEPR Discussion Paper No. 191*, CAEPR, ANU, Canberra.

—— and Pollack, D. P. 1998. 'Native Title compensation: Historic and policy perspectives for an effective and fair regime', *CAEPR Discussion Paper No. 152*, CAEPR, ANU, Canberra.

—— and Rowse, T. 2005. 'Indigenous affairs', in P. Saunders and J. Walter (eds), *Ideas and Influence: Social Science and Public Policy in Australia*, UNSW Press, Sydney.

—— and Sanders, W. 1991. 'From exclusion to dependence: Aborigines and the welfare state in Australia', *CAEPR Discussion Paper No. 1*, CAEPR, ANU, Canberra.

—— and Smith, D. E. 1994. 'The economic impact of mining moneys: The Nabarlek case, Western Arnhem Land', *CAEPR Discussion Paper No. 63*, CAEPR, ANU, Canberra.

——and Smith, D. E. 1999. 'The Ngurratjuta Aboriginal Corporation: A model for understanding Northern Territory royalty associations', *CAEPR Discussion Paper No. 185*, CAEPR, ANU, Canberra.

Anderson, I. 2004. 'The Framework Agreements: Intergovernmental agreements and Aboriginal and Torres Strait Islander health', in M. Langton, M. Tehan, L. Palmer and K. Shain (eds), *Honour Among Nations?: Treaties and Agreements with Indigenous People*, Melbourne University Press, Carlton.

Anderson, R. B. 2002. 'Entrepreneurship and Aboriginal Canadians: A case study in economic development', *Journal of Developmental Entrepreneurship,* 7 (1): 45–65.

Aristotle. 350BC. *Nicomachean Ethics,* Book 1, Translated by W. D. Ross, The Internet Classics Archive, viewed 26 June 2008, http://classics.mit.edu/Aristotle/nicomachaen.1.i.html

Armstrong, R., Morrison, J. and Yu, P. 2005. 'Indigenous land and sea management and sustainable business development in Northern Australia', *NAILSMA Discussion Paper,* North Australian Land and Sea Management Alliance, Darwin.

Arthur, W. S. 1999. 'What's new? The 1997 parliamentary inquiry into indigenous business', *CAEPR Discussion Paper No. 177,* CAEPR, ANU, Canberra.

Attwood, B. 2003. *Rights for Aborigines,* Allen and Unwin, Sydney.

Austin-Broos, D. 2003. 'Places, practices, and things: The articulation of Arrernte kinship with welfare and work', *American Ethnologist,* 30 (1): 118–35.

Austin-Broos, D. and Macdonald, G. (eds) 2005. *Culture, Economy and Governance in Aboriginal Australia,* Sydney University Press, Sydney.

Australian Bureau of Statistics (ABS) 2002. *Statistical Geography: Volume 2— Census Geographic Areas, Australia, 2001,* cat. no. 2905.0, ABS, Canberra.

—— 2004. *Australian Social Trends 2004,* cat. no. 4102, ABS, Canberra.

—— 2006a. *Statistical Geography: Volume 1—Australian Standard Geographical Classification (ASGC), July 2006,* cat. no. 1216.0, ABS, Canberra.

—— 2006b. *Housing and Infrastructure in Aboriginal and Torres Strait Islander Communities: Australia 2006,* cat. no. 4710.0, ABS, Canberra.

—— 2007. *Census of Population and Housing—Undercount,* cat. no. 2940.0, ABS, Canberra.

Australian Institute of Aboriginal Studies 1984. *Aborigines and Uranium: Consolidated Report on the Social Impact of Uranium Mining on the Aborigines of the Northern Territory,* AGPS, Canberra.

Australian Minerals and Energy Environment Foundation 2002. Facing the Future: The Report of the MMSD Australia Project, Mining Minerals and Sustainable Development Project, Melbourne.

Australian Non-Government Organisations 1998. 'Principles for the conduct of company operations within the minerals industry', Australian Asia-Pacific Mining Network, Sydney.

Auty, R. 1993. *Sustaining Development in Mineral Economies: The Resource Curse Thesis*, Routledge, London.

Ballard, C. and Banks, G. 2003. 'Resource Wars: The anthropology of mining', *Annual Reviews in Anthropology*, 32: 287–313.

Balsamo, F. and Calma, T. 2007. 'Is economic development possible on Indigenous land?' Paper presented to the *Collaborative Indigenous Policy IQPC 6th Annual Conference*, 1–2 May, Brisbane.

Banerjee, B. 2000. 'Whose land is it anyway? National interest, indigenous stakeholders, and colonial discourses', *Organization & Environment*, 13 (1): 3–39.

—— 2001. 'Corporate citizenship and Indigenous stakeholders—Exploring a new dynamic of organisational-stakeholder relationships', *The Journal of Corporate Citizenship*, 1: 39–55.

Bang, H. P. and Bech Dyrbery, T. 2000. 'Governance, self-representation and democratic imagination', in M. Saward (ed.), *Democratic Innovation*, Routledge, London.

Barber, B. 1984. *Strong Democracy: Participatory Politics for a New Age*, University of California Press, Berkeley.

Barker, T. 2006. 'Employment outcomes for Aboriginal people: An exploration of experiences and challenges in the Australian minerals industry', *CSRM Research Paper No. 6*, Centre for Social Responsibility in Mining, University of Queensland, Brisbane.

—— and Brereton, D. 2004. 'Aboriginal employment at Century Mine', *CSRM Research Paper No. 3*, Centre for Social Responsibility in Mining, University of Queensland, Brisbane.

—— and —— 2005. 'Survey of local Aboriginal people formerly employed at Century Mine', *CSRM Research Paper No. 4*, Centre for Social Responsibility in Mining, University of Queensland, Brisbane.

Batty, P. 2005. 'Private politics, public strategies: white advisers and their Aboriginal subjects', *Oceania*, 75 (3): 209–21.

Bell, M. 1992. *Demographic Projections and Forecasts in Australia: A Directory and Digest*, AGPS, Canberra.

Bendell, J. 2000a. 'Civil regulation: A new form of democratic governance for the global economy?', in J. Bendell (ed.), *Terms for Endearment: Business, NGOs and Sustainable Development*, Greenleaf Publishing, Sheffield.

—— 2000b. 'Talking for change? Reflections on effective stakeholder dialogue', Paper, New Academy of Business Innovation Network, London, viewed 4 June 2003,

<www.new-academy.ac.uk/research/publications/document/talkingforchange.pdf>.

Birckhead, J. 1999. 'Brief encounters: Doing rapid ethnography in Aboriginal Australia', in S. Toussaint and J. Taylor (eds), *Applied Anthropology in Australasia*, University of Western Australia Press, Perth.

Blaser, M. 2004. 'Life projects: Indigenous people's agency and development', in M. Blaser, H. Feit and G. McRae (eds), *In the Way of Development: Indigenous Peoples, Life Projects and Globalization*, Zed Books, London.

Blowes, R. and Trigger, D. 1998. 'North Queensland case study: The Century Mine Agreement', in M. Edmunds (ed.), *Regional Agreements Key Issues in Australia, Vol. 1 Summaries*, Native Title Research Unit, AIATSIS, Canberra.

Bomford, M. and Caughley, J. (eds) 1996. *Sustainable Use of Wildlife by Aboriginal Peoples and Torres Strait Islanders*, Bureau of Resource Sciences, AGPS, Canberra.

Bourdieu, P. 1977. *Outline of a Theory of Practice* (trans. R. Nice), Cambridge University Press, Cambridge.

—— 1986. 'The forms of capital', in J. Richardson (ed.), *Handbook of Theory and Research for the Sociology of Education*, Greenwood Press, New York.

Bradshaw, E. 1998. 'Trains, planes and automobiles: Iron ore mining and archaeology', Paper presented to the *Australian Archaeological Association Annual Conference*, 10–12 December, Valla Beach, New South Wales.

Brennan, F. 1998. *The Wik Debate: Its Impact on Aborigines, Pastoralists and Miners*, UNSW Press, Sydney.

Brereton, D. 2003. 'Promoting sustainable development in the minerals industry: A multi-disciplinary approach', Edited version of a keynote presentation to the *Annual Conference of the Australasian Association for Engineering Education*, September, Melbourne, viewed 10 December 2008, <http://www.csrm.uq.edu.au/docs/SD_Minerals.pdf>.

Brereton, D. and Parmenter, J. 2008. 'Indigenous employment in the Australian mining industry', *Journal of Energy and Natural Resources Law*, 26 (1): 66–90.

Bridge, G. 2004. 'Contested terrain: Mining and the environment', *Annual Review of Environment and Resources*, 29: 205–59.

Calma, T. 2005. 'Overcoming Indigenous disadvantage key indicators report 2005: A human rights perspective', Paper presented to *Workshop on the Overcoming Indigenous Disadvantage Report*, Australian Human Rights and Equal Opportunity Commission, 16 September, Sydney.

Carpentaria Land Council Aboriginal Corporation 2004. 'Response of the CLCAC on behalf of the Waanyi Native Title Group to the Draft report by the State of Queensland, PCML and GADC and David Martin', Unpublished report prepared for the purposes of the First 5 Year Review of the 1997 Century Agreement, Carpentaria Land Council Aboriginal Corporation, Mt Isa.

Clifford, L. (Chief Executive Rio Tinto) 2001. 'Building on our strengths', Paper presented at the Securities Institute of Australia, 22 March, Sydney, viewed 24 April 2008, <http://www.riotinto.com/documents/Media/Speeches/RLC_SIA_speech_22.3.01.pdf>.

Cohen, A. P. 1985. *The Symbolic Construction of Community*, Ellis Horwood, Chichester.

—— 1993. 'Culture as identity: An anthropologist's view', *New Literary History*, 24 (1): 195–209.

Coleman, J. 1988. 'Social capital in the creation of human capital', *American Journal of Sociology 94 (Supplement)*: S95–S120.

Commonwealth of Australia 1991. *Royal Commission into Aboriginal Deaths in Custody: National Report*, Commissioner E. Johnston, AGPS, Canberra.

—— 2000a. *Health is Life: Report on the Inquiry into Indigenous Health*, House of Representatives Standing Committee on Family and Community Affairs, Canberra.

—— 2000b. *Draft Report of the Indigenous Funding Inquiry*, Commonwealth Grants Commission Discussion Paper IFI 2000/2, Commonwealth Grants Commission, Canberra.

—— 2004. *Many Ways Forward: Report of the Inquiry into Capacity Building and Service Delivery in Indigenous Communities*, House of Representatives Standing Committee on Aboriginal and Torres Strait Islander Affairs, Canberra.

—— 2005. *After ATSIC—Life in the Mainstream?*, Senate Select Committee on the Administration of Indigenous Affairs, Canberra.

Community Economies Collective 2001. 'Imagining and enacting noncapitalist futures', *Socialist Review*, 28 (3/4): 93–135.

Cook, F. 1997. 'Current Issues Brief 19 1996–97: Sale of the Century Zinc Project', Department of the Parliamentary Library, Canberra, viewed 19 November 2002, <www.aph.gov.au/library/pubs/cib/1996-97097cib19.htm>.

Corbett, T. and O'Faircheallaigh, C. 2006. 'Unmasking the politics of Native Title: The National native Title tribunal's Application of the NTA's arbitration provisions', *UWA Law Review*, 33: 153–77.

Corporations and Markets Advisory Committee 2006. The Social Responsibility of Corporations, Report to the Australia Government, Canberra, viewed 23 May 2007, <http://www.camac.gov.au/camac/camac.nsf/byHeadline/PDFFinal+Reports+2006/$file/CSR_Report.pdf >.

Council of Australian Governments 2008. Council of Australian Governments Communiqué, 29 November 2008, viewed 24 December 2008, <http://www.coag.gov.au/coag_meeting_outcomes/2008-11-29/index.cfm#indigenous>.

Cousins, D. and Nieuwenhuysen, J. 1984. *Aboriginals and the Mining Industry: Case Studies of the Australian Experience*, George Allen and Unwin, Sydney.

Cragg, W. and Greenbaum, A. 2002. 'Reasoning about responsibilities: Mining company managers on what stakeholders are owed', *Journal of Business Ethics*, 39 (3): 319–35.

Crooke, P., Harvey, B. and Langton, M. 2006. 'Implementing and monitoring Indigenous Land Use Agreements: The Western Cape Communities Co-existence Agreement' in M. Langton, O. Mazel, L. Palmer, K. Shain and M. Tehan (eds), *Settling with Indigenous People: Modern Treaty and Agreement Making*, The Federation Press, Sydney.

Crush, J. (ed.) 1997. *Power of Development,* Routledge, London.

Cultural Survival 2001. 'Mining Indigenous lands: Can impacts and benefits be reconciled?', *Cultural Survival Quarterly*, 25: 1.

Cunningham, J. 1998. 'Implications of changing Indigenous population estimates for monitoring health trends', *Australasian Epidemiologist*, 5 (1): 6–8.

Cusack, B. 2001. 'Mining and communities—working together', Paper presented at the *Mining and Communities Conference—Working Together*, 25 September, Bendigo.

Daffen, P. 1995. *Working Together: A Review of DAS Strategies for Aboriginal and Torres Strait Islander Development*, AGPS, Canberra.

Dahl, R. 1985. 'Can international organizations be democratic?', in I. Shapiro and C. Hacker-Cordon (eds), *Democracy's Edges*, Cambridge University Press, Cambridge.

Davis, L. 1995. 'New directions for CRA', Speech delivered to the Securities Institute Australia, March, Sydney.

—— 2002. 'The corporate sector and social and community commitment', *Journal of Indigenous Policy*, 2: 37–41.

Davis, R. 2005. 'Identity and economy in Aboriginal pastoralism', in L. Taylor, G. K. Ward, G. Henderson, R. Davis and A. Wallis (eds), *The Power of*

Knowledge, The Resonance of Tradition, Aboriginal Studies Press, Canberra.

Davis, S. L. and Prescott, J. R. V. 1992. *Aboriginal Frontiers and Boundaries in Australia*, Melbourne University Press, Melbourne.

Dean, M. 1999. *Governmentality: Power and Rule in Modern Society*, Sage Publications, London.

—— 2002. 'Liberal government and authoritarianism', *Economy and Society* 31: 37–61.

de Haan, L. and Zoomers, A. 2005. 'Exploring the frontiers of livelihoods research', *Development and Change*, 36 (1): 27–47.

Department of Industry Tourism and Resources 2006. Working in Partnership Program, Australian Government, Canberra, viewed 5 February 2007, <http://www.industry.gov.au/content/ sitemap.cfm?objectID=BE9164E7-AD58-9A9B-3360C00A1FF46F10>.

de Soto, H. 2000. *The Mystery of Capital: Why Capitalism Triumphs in the West and Fails Everywhere Else*, Bantam, Ealing.

de Tocqueville, A. 1945. *Democracy in America*, New York, Random House.

Dixon, R. 1990. 'Aborigines as purposive actors or passive victims?: An account of the Argyle events by some of the Aboriginal participants', in R. A. Dixon and M. C. Dillon (eds), *Aborigines and Diamond Mining: the Politics of Resource Development in the East Kimberley Western Australia*, University of Western Australia Press, Nedlands.

Dodson, M. and Smith, D. 2003. 'Governance for sustainable development: Strategic issues and principles for Indigenous Australian communities', *CAEPR Discussion Paper No. 250*, CAEPR, ANU, Canberra.

Doohan, K. 2003. 'Argyle Diamonds and Wirnan: A process of incorporation and (re)negotiating relationships in the East Kimberley', Presentation to Department of Human Geography, Macquarie University, September, viewed 10 December 2008, <http://dlc.dlib.indiana.edu/archive/00001209/00/DoohanBris03.pdf>.

—— 2008. *Making Things Come Good: Relations between Aborigines and Miners at Argyle*, Backroom Press, Broome.

Downing, T., Garcia-Downing, C., Moles, J. and McIntosh, I. 2003. 'Mining and indigenous peoples: Stakeholder strategies and tactics', in *Finding Common Ground: Indigenous People and their Association with the Mining Sector*, International Institute for Environment and Development and World Business Council for Sustainable Development, London, viewed

5 May 2008,
<http://www.iied.org/mmsd/mmsd_pdfs/commonground.pdf>.

Dryzek, J. 1996a. *Democracy in Capitalist Times*, Oxford University Press, New York.

—— 1996b. 'Political inclusion and the dynamics of democratization', *The American Political Science Review*, 90 (1): 475–87.

—— 2000a. *Deliberative Democracy and Beyond: Liberals, Critics, Contestations*, Oxford University Press, Oxford.

—— 2000b. 'Discursive democracy vs liberal constitutionalism', in M. Saward (ed.), *Democratic Innovation: Deliberation, Representation and Association*, Routledge, London.

Dubbink, W. 2005. 'Democracy and private discretion in business', *Business Ethics Quarterly*, 15 (1): 37–66.

Durie, M. 2005. *Nga Tai Matatu Tides of Maori Endurance*, Oxford University Press, Melbourne.

Earth Tech 2005. *Southern Gulf Catchments Natural Management Plan: The Assets, Threats and Targets of the Region, Book 4*, Southern Gulf Catchments, Natural Heritage Trust, Mt Isa.

Eckersley, R. 2000. 'Deliberative democracy, ecological representation and risk: Towards a democracy of the affected', in M. Saward (ed.), *Democratic Innovation:Deliberation, Representation and Association*, Routledge, London.

Edmunds, M. 1989. *They Get Heaps: A Study of Attitudes in Roebourne Western Australia*, The Institute Report Series, Aboriginal Studies Press, Canberra.

Eggleston, P. (Group Coordinator for Sustainable Development Rio Tinto) 2002. 'Gaining Aboriginal community support for new mine development and making a contribution to sustainable development', Paper presented at the *Energy and Resources Law Conference*, 14–19 April, Edinburgh.

Energy Resources of Australia Ltd (ERA) 2006. Energy Resources of Australia Ltd website, viewed 6 January 2008, <http://www.energyres.com.au>.

Escobar, A. 1995. *Encountering Development: The Making and Unmaking of the Third World*, Princeton University Press, Princeton.

Esteva, G. 2005. 'Development', in W. Sachs (ed.), *The Development Dictionary*, Zed Books Ltd., London.

Etzioni-Halevy, E. 2003. 'Network governance as a challenge to democratic elite theory', Paper presented at the *Conference on Democratic Network Governance*, 22–23 May, Copenhagen.

Ferguson, J. 1994. *The Anti-Politics Machine: 'Development', Depoliticization, and Bureaucratic Power in Lesotho*, University of Minnesota Press, Minneapolis.

Fine, B. 2003. 'Review Essay. Social capital: The World Bank's fungible friend', *Journal of Agrarian Change*, (3) 4: 586–603.

Finlayson, J. (Australian Collaboration) 2007a. *Maps to Success: Successful Strategies in Aboriginal Organisations*, AIATSIS, Canberra.

—— 2007b. *Organising for Success: Policy Report, Successful Strategies in Aboriginal Organisations*, AIATSIS, Canberra.

Fliedner, I. (Director Corporate Affairs BP Amoco Australasia) 2000. 'The shifting boundaries of the private sector—Are we moving towards Corporate Citizenship?' Paper presented at the *The Second National Conference on Corporate Citizenship*, 7 April, Canberra.

Flucker, D. 2003a. Discussion Paper: Undertaking a Regional Process to Address Regional Issues, Unpublished report, Carpentaria Land Council Aboriginal Corporation on behalf of the Waanyi Nation Aboriginal Corporation, Mt Isa.

—— 2003b. Waanyi Discussion Paper, Waanyi Position on Century Mine Agreement: 60 Percent Impact Equals 60 Percent Benefits, Unpublished report, Carpentaria Land Council Aboriginal Corporation on behalf of the Waanyi Nation Aboriginal Corporation, Mt Isa.

Folds, R. 2001. *Crossed Purposes: The Pintupi and Australia's Indigenous Policy*, UNSW Press, Sydney.

Foley, D. 2003. 'An examination of Aboriginal Australian entrepreneurs', *Journal of Developmental Entrepreneurship,* (8) 2: 131–51.

—— 2006. 'Aboriginal Australian entreprenuers: Not all community organisations, not all in the Outback', *CAEPR Discussion Paper No. 279,* CAEPR, ANU, Canberra.

Fombrun, C. 1997. 'Three pillars of corporate citizenship: Ethics, social benefit, profitability', in N. Tichy, A. McGill and L.St Clair (ed.), *Corporate Global Citizenship,* The New Lexington Press, San Francisco.

Foucault, M. 1991. 'Governmentality', in G. Burchell, C. Gordon, and P. Miller (eds), *The Foucault Effect: Studies in Governmentality*, University of Chicago Press, Chicago.

Fox, C. and Miller, H. 1995. *Postmodern Public Administration: Towards Discourse,* Sage, Thousand Oaks, California.

Fukuda-Parr, S. Lopes, C. and Malik, K. 2002. *Capacity for Development: New Solutions to Old Problems*, Earthscan and UNDP, London.

Fox, R. W., Kelleher, G. H. and Kerr, C. B. 1977. *Ranger Uranium Environmental Inquiry, Second Report*, AGPS, Sydney.

Gerritsen, R. 1982. 'Blackfellas and whitefellas', in P. Loveday (ed.), *Service Delivery to Remote Communities*, North Australia Research Unit Monograph, ANU, Darwin.

Gibson-Graham, J. K. 2002. 'Beyond global vs. local: Economic politics outside the binary frame', in A. Herod and M. Wrights (eds), *Geographies of Power: Placing Scale,* Blackwell Publishers, Oxford.

——— 2005. 'Surplus possibilities: Post-development and community economies', *Singapore Journal of Tropical Geography*, 26 (1): 4–26.

Glaskin, K. 2003. 'Native Title and the "bundle of rights" model: Implications for the recognition of Aboriginal relations to country', *Anthropological Forum*, 13 (1): 67–88.

——— 2007. 'Claim, culture and effect: property relations and the native title process', in B. Smith and F. Morphy (eds), *The Social Effects of Native Title: Recognition, Translation, Coexistence*, CAEPR Research Monograph No. 27, ANU E Press, Canberrra.

Goghill, K. 2002. 'Governance for uncertain times', Paper presented at the Judge Institute of Management Studies Cambridge University, 18 June, Cambridge.

Gray, W. J. 1977. 'Decentralisation trends in Arnhem Land', in R. M. Berndt (ed.), *Aborigines and Change: Australia in the '70s,* Aboriginal Institute of Aboriginal Studies, Canberra.

Gray, I. and Lawrence, G. 2001. *A Future for Regional Australia: Escaping Global Misfortune*, Cambridge University Press, Melbourne.

Gray, W. and Sanders, W. G. 2006. 'Views from the top of the "quiet revolution": Secretarial perspectives on the new arrangements in Indigenous affairs', *CAEPR Discussion Paper No. 282*, CAEPR, ANU, Canberra.

Griffiths, T. 2000. Sustainability of Wildlife Use for Subsistence in the Maningrida Region, Progress report to Natural Heritage Trust, Centre for Tropical Wildlife Management, Northern Territory University, Darwin.

Gulf Communities Agreement 1997. 'Agreement between the Waanyi, Mingginda, Gkutharn and Kukatj peoples and State of Queensland and Century Zinc Limited, An Agreement under the right to negotiate provisions of the Native Title Act 1993 in relation to the Century Zinc Project', unpublished.

Gumala Aboriginal Corporation 2005. 'Karijini Eco Retreat', Gumala Aboriginal Corporation, Perth, viewed 10 December 2008, <http://www.gumala.com.au/eco-tourism.htm>.

Gunjeihmi Aboriginal Corporation 2001. Submission by the Mirrar Aboriginal People, Kakadu, Australia to The Office of the High Commissioner for Human Rights Workshop on 'Indigenous peoples, private sector natural resource, energy and mining companies and human rights', Unpublished submission, Jabiru.

—— 2003. November 2003 Royalty Equivalent Budget, Gunjeihmi Aboriginal Corporation, Jabiru.

—— 2006. The History of Binninj Opposition to Uranium Mining, Mirarr website, Jabiru, viewed 7 January 2009, <http://www.mirarr.net/history.html>.

Gunningham, N., Kagan, R. and Thornton, D. 2002. 'Social licence and environmental protection: Why businesses go beyond compliance', Discussion Paper, Centre for Analysis of Risk and Regulation, London School of Economics and Political Science, London.

Gusfield, J. R. 1975. *Community: A Critical Response*, Harper Colophon, New York.

Hall, J. and Driver, M. 2002. Queensland Support for Training and Employment Through the Gulf Communities Agreement and Century Mine, Unpublished report, Queensland Department of Employment and Training, Brisbane.

Hamersley Iron 2000 (updated). 'Hamersley's Five Year Communities Plan 1998–2002', Internal company report, Perth.

Hamilton, C. 2001. *Running from the Storm: The Development of Climate Change Policy in Australia*, UNSW Press, Sydney.

—— and Maddison, S. (eds) 2007. *Silencing Dissent: How the Australian Government is Controlling Public Opinion and Stifling Debate,* Allen & Unwin, Sydney.

Harvey, B. 2002a. 'New competencies in mining: Rio Tinto's experience', Paper presented at the *Council of Mining and Metallurgical Congress*, 27–28 May, Cairns.

—— 2002b. 'Sociology before geology: The development of social competencies in Rio Tinto', Paper presented at the *Sustainable Development Conference*, MCA, November, viewed 5 May 2008, <http://www.riotinto.com/documents/MediaSpeeches/BEHSociology_before_Geology.pdf >.

—— 2004. 'Rio Tinto's agreement making in Australia in a context of globalisation', in M. Langton, M. Tehan, L. Palmer, and K. Shain (eds),

Honour Among Nations? Treaties and Agreements with Indigenous Peoples, Melbourne University Press, Melbourne.

—— and Brereton, D. 2005. 'Emerging models of community engagement in the Australian minerals industry', Paper presented at the *International Conference on Engaging Communities*, 14–17 August, Brisbane, viewed 5 May 2008, <http://www.riotinto.com/documents/MediaSpeeches/ UN_Conference_on_Community_Engagement_BH_150805.pdf>.

—— and J. Gawler, 2003. 'Aboriginal employment diversity in Rio Tinto', *International Journal of Diversity in Organisations, Communities and Nations* (3): 197–209.

Harvey, D. 2007. *A Brief History of Neoliberalism*, Oxford University Press, Oxford.

Hawke, S. and Gallagher, M. 1989. *Noonkanbah: Whose Land, Whose Law*, Fremantle Arts Centre Press, Fremantle.

Haq, M. n.d. 'The Human Development concept', Human Development Reports, United National Development Program, viewed 10 December 2008, <http://hdr.undp.org/en/humandev/>.

Held, D. 1987. *Models of Democracy,* Polity Press, Cambridge.

Hill, R. (ed.) 1999. *Mining and Indigenous Peoples: Case Studies*, The International Council on Metals and the Environment, Ottawa, viewed 5 May 2008, >http://www.icmm.com/library_pub_detail.php?rcd=11>.

——, Golson, K., Lowe, P., Mann, M., Hayes, S. and Blackwood, J. 2005. *Kimberley Appropriate Economies Roundtable Forum Proceedings,* Kimberley Land Council, Environs Kimberley and Australian Conservation Foundation, Cairns.

Hindle, K. and Lansdowne, M. 2005. 'Brave spirits on new paths: Toward a globally relevant paradigm of Indigenous entrepreneurship research', *Journal of Small Business and Entrepreneurship*: *Special issue on Indigenous Entrepreneurship*, 18 (2): 131–41.

Hinkson, M. and B. Smith 2005. 'Introduction: Conceptual moves towards an intercultural analysis', *Oceania*, 75 (3): 157–66.

Hirst, P. 1994. *Associative Democracy,* Polity Press, Cambridge.

Hobart, M. (ed.) 1993. *An Anthropological Critique of Development: The Growth of Ignorance*, Routledge, London.

Hoffmeister, C. 2002. Review of the Gumala Foundations: Final Report to the Trustee Gumala Investments Pty Ltd., Unpublished report, Prime Focus, Perth.

Holcombe, S. 2004a. 'Early Indigenous engagement with mining in the Pilbara: Lessons from a historical perspective', *CAEPR Working Paper No. 24* [revised August 2006], CAEPR, ANU, Canberra.

—— 2004b. 'The sentimental journey: a site of belonging. A case study from central Australia', *The Australian Journal of Anthropology*, 15: 163–84.

—— 2006. '"Community Benefit Packages": Development's encounter with pluralism in the case of the mining industry', in T. Lea, E. Kowal and G. Cowlishaw (eds), *Moving Anthropology: Critical Indigenous Studies*, Charles Darwin University Press, Darwin.

Holme, Lord of Cheltenham (Advisor to the Chairman of Rio Tinto) 1999. 'Responsible business engagement with society', *BCA Papers,* Business Council of Australia, Melbourne.

Holmes, J. H. 2002. 'Diversity and change in Australia's rangelands: A post-productivist transition with a difference?' *Transactions of the Institute of British Geographers*, 27 (3): 362–84.

Hooke, M. 2007. National Press Club Address, 4 April, Canberra.

House of Representatives Standing Committee on Aboriginal and Torres Strait Islander Affairs 1999. *Unlocking the Future: The Report of the Inquiry into the Reeves Review of the Aboriginal Land Rights (Northern Territory) Act 1976*, Parliament of the Commonwealth of Australia, Canberra.

Howitt, R. 1998. 'Recognition, respect and reconciliation: Steps towards decolonisation?', *Australian Aboriginal Studies,* 2: 28–34.

—— 2001. *Rethinking Resource Management: Justice, Sustainability and Indigenous Peoples*, Routledge, London.

——, Connell, J. and Hirsch, P. 1996. 'Resources, nations and indigenous peoples', in R. Howitt with J. Connell and P. Hirsch (eds), *Resources, Nations and Indigenous Peoples: Case Studies from Australasia, Melanesia and Southeast Asia*, Oxford University Press, Melbourne.

Hunt, J. and D. Smith 2006a. 'Building Aboriginal community governance in Australia: Preliminary research findings', *CAEPR Working Paper No. 31*, CAEPR, ANU, Canberra.

—— and —— 2006b. 'Ten key messages from the preliminary findings of the Indigenous Community Governance Project 2005', Summary document associated with *CAEPR Working Paper No. 31*, CAEPR, ANU, Canberra.

—— and —— 2007. 'Aboriginal Community Governance Project: Year two research findings', *CAEPR Working Paper No. 36*, CAEPR, ANU, Canberra.

——, ——, Garling, S. and W. Sanders (eds) 2008. *Contested Governance: Culture, Power and Institutions in Indigenous Australia*, CAEPR Research Monograph No. 29, ANU E Press, Canberra.

Hunter, B. 1999. 'Three nations, not one: Indigenous and other Australian poverty', *CAEPR Working Paper No. 1*, CAEPR, ANU, Canberra.

——2004. 'Taming the social capital Hydra?', *Learning Communities: International Journal of Learning in Social Contexts*, 2: 19–51.

——2006. 'Revisiting the poverty wars: Income status and financial stress among indigenous Australians', in B.H. Hunter (ed.), *Assessing the Evidence on Indigenous Socioeconomic Outcomes: A Focus on the 2002 NATSISS*, CAEPR Research Monograph No. 26, ANU E Press, Canberra.

Hyndman, D. 1994. *Ancestral Rain Forests and the Mountain of Gold: Indigenous Peoples and Mining in New Guinea*, Westview, Boulder.

Indigenous Support Services and ACIL Consulting (ISS/ACIL) 2001. Agreements Between Mining Companies and Indigenous Communities: A Report to the Australian Minerals and Energy Environment Foundation, viewed 5 May 2008, <http://www.natural-resources.org/minerals/cd/ docs/mmsd/australia/finalreport/indigenous.pdf>.

Industry Commission 1991. *Mining and Mineral Processing in Australia, Report No. 7*, AGPS, Canberra.

Kakadu Region Social Impact Study 1997a. Community Action Plan. Report of the Study Advisory Group, Supervising Scientist, Canberra.

—— 1997b. *Report of the Aboriginal Project Committee*, Office of the Supervising Scientist, Canberra.

—— 1997c. *Report of the Project Advisory Committee*, Office of the Supervising Scientist, Canberra.

Kapelus, P. 2002. 'Mining, corporate social responsibility and the "community": The case of Rio Tinto, Richards Bay Minerals and the Mbonambi', *Journal of Business Ethics,* 39 (3): 275–96.

Karl, T. L. 1997. *The Paradox of Plenty: Oil Booms and Petro States*, University of California Press, Berkley.

Kemp, D., Boele, R. and Brereton, D. 2006. 'Community relations management systems in the minerals industry: combining conventional and stakeholder-driven approaches', *International Journal of Sustainable Development*, 9 (4): 390–403.

Keskinen, A., Aaltonen, M. and Mitleton-Kelly, E. 2003. *Organisational Complexity*, FFRC Publications 6/2003, Finland Futures Research Centre, Helsinki.

Kesteven, S. 1983. 'The effects on Aboriginal communities of monies paid out under the Ranger and Nabarlek Agreements', in N. Peterson and M. Langton (eds), *Aborigines, Land and Land Rights*, Australian Institute of Aboriginal Studies, Canberra.

—— 1986. 'The project to monitor the social impact of uranium mining on Indigenous communities in the Northern Territory', *Australian Indigenous Studies*, 1986/1: 43–5.

King, D. 2000. 'Stakeholders and Spin Doctors: The politicisation of corporate reputations', *Hawke Institute Working Paper Series No. 5*, Hawke Institute, University of South Australia, Magill.

Kirsch, S. 2006. *Reverse Anthropology: Indigenous Analysis of Social and Environmental Relations in New Guinea,* Stanford University Press, Stanford, Ca.

Klausen, J. E. and Sweeting, D. 2003. 'Legitimacy, citizen participation, and community involvement in governance', Paper presented at the *Conference on Democratic Network Governance*, 22–23 May, Copenhagen.

Kolig, E. 2005. 'The politics of Indigenous—or ingenious—tradition: Some thoughts on the Australian and New Zealand situation', in T. Otto and P. Pedersen (eds), *Tradition and Agency: Tracing Cultural Continuity and Invention*, Aarhus University Press, Aarhus.

Korten, D. 2001. 'Predatory corporations', in G. Evans, J. Goodman and N. Lansbury (eds), *Moving Mountains: Communities Confront Mining and Globalisation,* Otford Press, Sydney.

Kreager, P. 1982. 'Demography in situ', *Population and Development Review*, 8 (2): 237–66.

Kymlicka, W. 1995. *Multi-Cultural Citizenship: A Liberal Theory of Minority Rights*, Clarendon Press, Oxford.

Lane, M. 2001. 'Moral responsibilities and popular acceptability: forms, forums for corporate engagements in localities', Paper presented at the *Workshop on Moral Rights and Corporate Responsibilities*, December, Centre for Applied Philosophy and Public Ethics, ANU, Canberra.

—— and Chase, A. 1996. 'Resource development on Cape York Peninsula: Marginalisation and denial of indigigenous perspectives', in R. Howiit, J. Connell and P. Hirsch (eds), *Resources, Nations and Indigenous Peoples*, Oxford University Press, Melbourne.

Langton, M., Tehan, M., Palmer, L. and Shain, K. (eds) 2004. *Honour Among Nations: Treaties and Agreements with Indigenous People*, Melbourne University Press, Melbourne.

Langton, M. and Mazel, O. 2008. 'Poverty in the midst of plenty: Aboriginal people, the 'Resource Curse' and Australia's mining boom', *Journal of Energy and Natural Resources Law*, 26 (1): 31–65.

——, ——, Palmer, L., Shain, K. and Tehan, M. (eds) 2006. *Settling with Indigenous People: Modern Treaty and Agreement-Making*, Federation Press, Sydney.

Lavelle, A. 2001. 'The mining industry's campaign against native title: Some explanations', *The Australian Journal of Political Science*, 36 (1): 101–22.

Lawrence, R. 2005. 'Governing Warlpiri subjects: Indigenous employment and training programs in the central Australian mining industry', *Geographical Research*, 43 (1): 40–8.

Levitus, R. I. 1991. 'The boundaries of Gagudju Association membership: Anthropology, Law and public policy', in J. Connell and R. Howitt (eds), *Mining and Indigenous Peoples in Australasia*, Sydney University Press, Sydney.

—— 1999. 'Local organisations and the purpose of money', in J.C. Altman, F. Morphy and T. Rowse (eds), *Land Rights at Risk? Evaluations of the Reeves Report*, CAEPR Research Monograph No.14, CAEPR, ANU, Canberra.

—— 2003. Sacredness and Consultation: An Interpretation of the Coronation Hill Dispute, PhD Thesis, The Australian National University, Canberra.

—— 2005. 'Land rights and local economies: The Gagudju Association and the mirage of collective self-determination', in D. Austin-Broos and G. Macdonald (eds), *Culture, Economy and Governance in Aboriginal Australia*, Sydney University Press, Sydney.

Li, T. M. 2007. *The Will to Improve: Governmentality, Development and the Practice of Politics*, Duke University Press, Durham and London.

Libby, R. 1989. *Hawke's Law: The Politics of Mining and Aboriginal Land Rights in Australia*, University of Western Australia Press, Perth.

McCausland, R. 2005a. *Negotiating Shared Responsibility Agreements: a Toolkit*, Ngiya Institute for Indigenous Law, Policy and Practice, University of Technology, Sydney.

—— 2005b. 'The "new mainstreaming" of Indigenous affairs', Briefing Paper for the Ngiya Institute of Indigenous Law, Policy and Practice, Jumbunna Indigenous House of Learning, University of Technology, Sydney.

Macdonald, G. 1997. '"Recognition and justice": The traditional/historical contradiction in New South Wales', in D. E. Smith and J. Finlayson (eds),

Fighting Over Country: Anthropological Perspectives, CAEPR Research Monograph No. 12, CAEPR, ANU, Canberra.

—— 2000. 'Economies and personhood: Demand sharing among the Wiradjuri of New South Wales', in G. W. Wenzel, G. Hovelsrud-Broda and N. Kishigami (eds), *The Social Economy of Sharing: Resource Allocation and Modern Hunter-Gatherers*, Senri Ethnological Studies No. 53, National Museum of Ethnology, Osaka.

—— 2004. *Two Steps Forward, Three Steps Back*, LhR Press, Canada Bay.

Macklin, J. The Hon. (Minister for Families, Housing, Community Services and Indigenous Affairs) 2008. 'Beyond Mabo: Native title and Closing the Gap', 2008 Mabo Lecture, Delivered at James Cook University, Townsville, 21 May, viewed 10 December 2008, <http://www.jennymacklin.fahcsia.gov.au/internet/jennymacklin.nsf/content/beyond_mabo_21may08.htm>.

McKenna, S. 1995. 'Assessing the relative allocative efficiency of the *Native Title Act 1993* and *Aboriginal Land Rights (Northern Territory) Act 1976*', *CAEPR Discussion Paper No. 79*, CAEPR, ANU, Canberra.

McLaren, D. 2002. 'Corporate engagement by "socially responsible" investors: A practical paradigm for stakeholder governance?', Essay, Judge Institute of Management, Cambridge.

Maddox, G. 1991. *Australian Democracy in Theory and Practice,* Longman Cheshire, South Melbourne.

Mander, J. and Tauli-Corpus, V. (eds) 2006. *Paradigm Wars: Indigenous Peoples' Resistance to Globalisation,* Sierra Club Books, San Francisco.

Mantziaris, C. and Martin, D. 2000. *Native Title Corporations: A Legal and Anthropological Analysis*, Federation Press, Sydney.

Martin, D. F. 1993. Autonomy and Relatedness: An Ethnography of Wik People of Aurukun, Western Cape York Peninsula, PhD thesis, ANU, Canberra.

—— 1995. 'Money, business and culture: Issues for Aboriginal economic policy', *CAEPR Discussion Paper No. 101*, CAEPR, ANU, Canberra.

—— 1997. 'The incorporation of "traditional" and "historical" interests in Native Title Representative Bodies', in D. E. Smith and J. Finlayson (eds), *Fighting Over Country: Anthropological Perspectives*, CAEPR Research Monograph No. 12, CAEPR, ANU, Canberra.

—— 1998a. 'Deal of the Century?—A case study from the Pasminco Century project', *Aboriginal Law Bulletin,* 4 (11): 4–7.

—— 1998b. 'Indigenous corporate structures for doing business with industry—a case study from the Century project', Paper presented at the *Doing*

Business with Aboriginal Communities Conference, 24–27 February, Alice Springs.

—— 2001. 'Is Welfare dependence 'Welfare Poison'? An assessment of Noel Pearson's proposals for Aboriginal welfare reform', *CAEPR Discussion Paper No. 213*, CAEPR, ANU, Canberra.

—— 2003. 'Rethinking the design of Aboriginal organisations: the need for strategic engagement', *CAEPR Discussion Paper No. 248*, CAEPR, ANU, Canberra.

—— 2004. 'Designing institutions in the native title recognition space', in S. Toussaint (ed.), *Crossing Boundaries: Cultural, Legal and Historical Issues in Native Title*, Melbourne University Press, Melbourne.

—— 2005a. 'Governance, cultural appropriateness, and accountability within the context of Indigenous self-determination', in D. Austin-Broos and G. Macdonald (eds), *Culture, Economy and Governance in Aboriginal Australia*, Sydney University Press, Sydney.

—— 2005b. 'Rethinking Aboriginal community governance: challenges for sustainable engagement', in P. Smyth, T. Reddel and A. Jones (eds), *Community and Local Governance in Australia*, UNSW Press, Sydney.

—— 2008a. 'Aboriginal sorcery and healing, and the alchemy of Aboriginal policy making', *Journal of the Anthropological Society of South Australia*, 33: 75–128.

—— 2008b. Expert's Affidavit to the Court on Background Socio-cultural Factors in the Case *R v KU, AAC, WY, PAG, KY, KZ, BBL, WZ & YC; ex parte A–G (Qld)* [2008] QCA 154 (the appeal in the matter of the 'Aurukun Nine'), June 2008.

—— and Finlayson, J. D. 1996. 'Linking accountability and self-determination in Aboriginal organisations', *CAEPR Discussion Paper No. 116*, CAEPR, ANU, Canberra.

——, Hondros, J. and Scambary, B. 2004. 'Enhancing Aboriginal social sustainability through agreements with resource developers', Paper presented to the *Inaugural Sustainable Development Conference*, MCA, October, Melbourne.

—— and Taylor, J. 1996. 'Ethnographic perspectives on the enumeration of Indigenous people in remote Australia', *Journal of the Australian Population Association*, 13 (1): 17–33.

Mays, S. 2003. 'Corporate sustainability—an investor perspective', Report to the Department of Environment and Heritage, Canberra.

McClelland, R. (Commonwealth Attorney-General) 2008. Speech at the *Negotiating Native Title Forum*, 29 February, Brisbane, viewed 10 December 2008, <http://www.attorneygeneral.gov.au/www/ministers/ RobertMc.nsf/Page/Speeches_2008_29February2008- NegotiatingNativeTitleForum>.

Merlan, F. 1998. *Caging the Rainbow: Place, Politics, and Aborigines in a North Australian Town*, University of Hawai'i Press, Honolulu.

—— 2005. 'Explorations towards intercultural accounts of socio-cultural reproduction and change', *Oceania*, 75 (3), 167–82.

Miller, H. 2003. 'Extra-formal democracy as pragmatic liberalism', Paper presented at the *Conference on Democratic Network Governance*, 22-23 May, Denmark.

Minerals Council of Australia (MCA) 2004. *Indigenous Relations Strategic Framework December 2004*, MCA, Canberra.

—— 2006. Minerals Council of Australia 2007–2008 Pre-budget Submission, December, viewed 13 March 2007, <http://www.minerals.org.au/__data/assets/pdf_file/18816/ FINAL_Pre-Budget_sub.pdf>.

—— 2007. *2007 Annual Report Minerals Council of Australia*, MCA, Canberra.

Mining Minerals and Sustainable Development 2002. *Facing the Future: The Report of the Mining Minerals and Sustainable Development Australia Project*, International Institute for Environment and Development, Melbourne.

Mitchell, R., Agle, B. and Wood, D. 1997. 'Toward a theory of stakeholder identification and salience: Defining the principle of who and what really counts', *The Academy of Management Review*, 22 (4): 853–86.

Moon, J. 2003. 'Corporate social responsibility and societal governance', *The Corporate Citizen*, (3) 1: 2–9.

Morgan, S., Kwaymullina, A. and Kwaymullina, B. 2006. 'Bulldozing Stonehenge: Fighting for cultural heritage in the wild wild west', *Indigenous Law Bulletin*, (6) 20: 6.

Morphy, F. 2002. 'When systems collide: The 2001 census at a Northern Territory outstation', in D. F. Martin, F. Morphy, W. G. Sanders, J. Taylor, *Making Sense of Census Data: Observation of the 2001 Enumeration in Remote Indigenous Australia*, CAEPR Research Monograph No. 22, ANU E Press, Canberra.

—— 2007. 'Uncontained subjects: "Population" and "household" in remote Aboriginal Australia'. *Journal of Population Research*, 24 (2): 163-84.

Morphy, H. 1999. 'The Reeves Report and the idea of the "region"', in J. C. Altman, F. Morphy and T. Rowse (eds), *Land Rights at Risk? Evaluations of the Reeves Report*, CAEPR Research Monograph No. 14, CAEPR, ANU, Canberra.

Murdock, S. H. and Ellis, D. R. 1991. *Applied Demography: An Introduction to Basic Concepts, Methods and Data*, Westview Press, Boulder, Colorado.

Murphy, D. and Bendell, J. 1999. *Partners in Time? Business, NGOs and Sustainable Development*, United Nations Research Institute for Social Development, Geneva.

Myers, F. 1986. Pintupi Country, *Pintupi Self: Sentiment, Place and Politics among Western Desert Aborigines*, Australian Institute of Aboriginal Studies, Canberra.

Native Title Payments Working Group 2008. Native Title Payments Working Group Report, Department of Families, Housing, Community Services and Indigenous Affairs, Canberra, viewed 24 December 2008, <http://www.fahcsia.gov.au/indigenous/native_title_wg_report/default.htm>.

Nederveen Pieterse, J. 1994. 'Globalisation as hybridisation', *International Sociology*, 9: 161–84.

Northern Land Council (NLC) 2006. *Celebrating Ten Years of Caring for Country: A Northern Land Council Initiative*, NLC, Darwin.

Northern Territory Government 2003. *Building Stronger Regions, Stronger Futures*, Northern Territory Department of Community Development, Sport and Cultural Affairs, Darwin.

Nussbaum, M. C. 2001. *Women and Human Development: The Capabilities Approach*, Cambridge University Press, Cambridge.

O'Brien, J. 2003. 'Canberra yellowcake: The politics of uranium and how Aboriginal land rights failed to Mirrar people', *Journal of Northern Territory History*, 14: 79–91.

O'Faircheallaigh, C. 1988. 'Uranium royalties and Aboriginal economic development', in D. Wade-Marshall and P. Loveday (eds), *Contemporary Issues in Development*, Australian National University North Australia Research Unit, Darwin.

—— 1996. 'Negotiating with resource companies: Issues and constraints for Aboriginal communities in Australia', in R. Howitt, J. Connell and P. Hirsch (eds), *Resources, Nations and Indigenous Peoples*, Oxford University Press, Melbourne.

—— 2000. 'Negotiating major project agreements: The Cape York model', *Research Discussion Paper 11*, AIATSIS, Canberra.

—— 2002a. 'Implementation: The forgotten dimension of agreement making in Australia and Canada', *Indigenous Law Bulletin*, 5 (20): 14–7.

—— 2002b. *A New Approach to Policy Evaluation: Mining and Indigenous People*, Ashgate Publishing, Aldershot.

—— 2003a. 'Financial models for agreements between indigenous peoples and mining companies', *Aboriginal Politics and Public Sector Management Research Paper No. 12,* Centre for Australian Public Sector Management, Griffith University, Brisbane.

—— 2003b. 'Implementing agreements between indigenous peoples and resource developers in Australia and Canada', *Aboriginal Politics and Public Sector Management Research Paper No. 13*, Centre for Australian Public Sector Management, Griffith University, Brisbane.

—— 2004a. 'Denying citizens their rights? Indigenous people, mining payments and service provision', *Australian Journal of Public Administration,* 63 (2): 45–50.

—— 2004b. 'Evaluating agreements between Indigenous peoples and resource developers', in M. Langton, M. Tehan, L. Palmer and K. Shain (eds), *Honour Among Nations?: Treaties and Agreements with Indigenous People*, Melbourne University Press, Carlton.

—— 2006. 'Aborigines, mining companies and the State in contemporary Australia: A new political economy or 'business as usual'?, *Australian Journal of Political Science*, 41 (1): 1–22.

Odd var Eriksen, E. 2000. 'The European Union's democratic deficit—A deliberative perspective', in M. Saward (ed.), *Democratic Innovation*, Routledge, London.

Office of Indigenous Policy Coordination, Department of Immigration and Multicultural and Indigenous Affairs 2004. New arrangements in Indigenous affairs, November update, Department of Immigration and Multicultural and Indigenous Affairs, Canberra.

Office of the Registrar of Indigenous Corporations 2008. 'Consolidated rule book', Documents for Ngumee-Ngu Aboriginal Corporation, viewed 10 December 2008, <http://www.oratsic.gov.au/document.aspx?concernID=103829>.

Office of the Registrar of Indigenous Corporations 1990. 'Corporation objects and rules', Documents for Karijini Aboriginal Corporation, viewed 10 December 2008, <http://www.oratsic.gov.au/document.aspx?concernID=101192>.

Olive, N. (ed.) 1997. *Karijini Mirlimirli: Aboriginal Histories from the Pilbara*, Fremantle Arts Centre Press, Freemantle.

Orlitzky, M. and Swanson, D. 2002. 'Value attunement: Toward a theory of socially responsible executivedecision-making', *Australian Journal of Management,* 27 Special Issue: 119–28.

Parker, C. 2002. *The Open Corporation,* Cambridge University Press, Cambridge.

Parliament of the Commonwealth of Australia 1999. Jabiluka: The Undermining of Process, Inquiry into the Jabiluka Uranium Mine Project, The Senate Environment, Communications, Information Technology and the Arts References Committee, Canberra.

—— 2006. *House of Representatives Aboriginal Land Rights (Northern Territory) Amendment Bill 2006, Explanatory Memorandum,* viewed 13 March 2007, <http://www.austlii.edu.au/au/legis/cth/bill_em/alrtab2006534/memo_0.html>.

Pasminco Century Mine 2001. Pasminco Century Mine—Agreement, Operations, Infrastructure, Community, Pasminco Century Mine, Garbutt.

—— 2003. 'Gulf Communities Agreement Employment Monthly Report—June 2003', Internal company report, Century Mine.

Pasminco n.d. 'Creating opportunities for Indigenous people and communities in north-west Queensland', viewed 19 November 2002, <www.isr.gov.au/resources/indigenouspartnerships/CaseStudies/pasminco/Pasminco.pdf >.

——, The State of Queensland and Gulf Aboriginal Development Corporation (GADC) 2002. Report on the Five-Year Review of the Century Mine Gulf Communities Agreement in Accordance With Clause 63 of the Agreement. Draft for comment August 2002, Unpublished, Brisbane.

Pateman, C. 1970. *Participation and Democratic Theory,* Cambridge University Press, Cambridge.

Pearson, N. 1999. 'Positive and negative welfare and Australia's indigenous communities', *Family Matters,* 54: 30–35.

—— 2000a. *Our Right to Take Responsibility,* Noel Pearson and Associates, Cairns.

—— 2000b. 'Passive welfare and the destruction of Indigenous society in Australia', in P. Saunders (ed.), *Reforming the Australian Welfare State,* Australian Institute of Family Studies, Melbourne.

—— 2004. 'Land is susceptible of ownership', in M. Langton, M. Tehan, L. Palmer and K. Shain (eds), *Honour Among Nations: Treaties and Agreements with Indigenous People,* Melbourne University Press, Melbourne.

—— and Kostakidis-Lianos, L. 2004. Building Indigenous Capital: Removing obstacles to participation in the real economy', *Australian Prospect* (1), Easter 2004, available at <http://www.australianprospect.com.au>.

Peredo, A. M., Anderson, R. B., Galbraith, C. S., Honig, B. and Dana, L. P. 2004 'Towards a theory of Aboriginal entrepreneurship', *International Journal Entrepreneurship and Small Business*, 1 (1,2): 1–20.

Peters-Little, F. 2000. 'The community game: Aboriginal self-definition at the local level', *AIATSIS Research Discussion Paper No. 10*, Aboriginal Studies Press, Canberra.

Peterson, N. 1993. 'Demand sharing: Reciprocity and the pressure for generosity among foragers', *American Anthropologist* 95 (4): 860–74.

—— 1996. 'Cultural issues', in J.C. Altman and J. Taylor (eds), *The 1994 National Aboriginal and Torres Strait Islander Survey: Findings and Future Prospects*, CAEPR Research Monograph No. 11, CAEPR, ANU, Canberra.

—— 1997. 'Demand sharing: Sociobiology and the pressure for generosity among foragers', in F. Merlan, J. Morton and A. Rumsey (eds), *Scholar and Sceptic: Australian Aboriginal Studies in Honour of L.R. Hiatt*, Aboriginal Studies Press, Canberra.

—— 2005. 'What can pre-colonial and frontier economies tell us about engagement with the real economy? Indigenous life projects and the conditions of development', in D. Austin-Broos and G. Macdonald (eds), *Culture, Economy and Governance in Aboriginal Australia*, Sydney University Press, Sydney.

Phibbs, P. 1989. 'Demographic-economic impact forecasting in non-metropolitan regions: An Australian example', in P. Congdon and P. Batey (eds), *Advances In Regional Demography: Information, Forecasts, Models*, Bellhaven Press, London.

Piggot, J. (Senior Vice President of Stern Stewart) 2002. 'Corporate Responsibility, is it worth it and for whom?', *The Ethical Corporation*, viewed 28 January 2003, <www.ethicalcorp.com/content_print.asp?ContentID=373>.

Pilbara Development Commission 2008. 'Pilbara Fund', viewed 11 December 2008, <http://www.pdc.wa.gov.au/funding--schemes/pilbara-fund-.aspx>.

Plumptre T. and Graham J. 1999. 'Governance and good governance: International and Aboriginal perspectives', Institute of Governance, Ottawa, viewed 11 December 2008, <http://www.iog.ca/publications/govgoodgov.pdf>.

Pollack, D. P. 2001. 'Indigenous land in Australia: A quantitative assessment of Indigenous landholdings in 2000', *CAEPR Discussion Paper No. 221*, CAEPR, ANU, Canberra.

Povinelli, E. A. 1993. *Labor's Lot: The Power, History and Culture of Aboriginal Action*, University of Chicago Press, Chicago.

—— 2002. *The Cunning of Recognition: Indigenous Alterities and the Making of Australian Multiculturalism*, Duke University Press, Durham, NC.

Power, M. 2003. 'Risk management and socially responsible corporations', Paper presented at the National Institute of Government and Law, Regulatory Institutions Network and the Australian National Centre for Audit and Assurance, 7 February, Canberra.

Pretes, M. 2005. Renewing the Wealth of Nations, PhD Thesis, Division of Society and the Environment, Research School of Pacific and Asian Studies, ANU.

Quiggin, J. 2005. 'Economic liberalism: Fall, revival and resistance', in P. Saunders and J. Walter (eds), *Ideas and Influence: Social Science and Public Policy in Australia*, UNSW Press, Sydney.

Rayner, M. 1997. *Rooting Democracy—Growing the Society We Want*, Allen and Unwin, St Leonards.

Redmond, A. 2007. 'Some initial effects of pursuing and achieving native title recognition in the northern Kimberley', in B. Smith and F. Morphy (eds), *The Social Effects of Native Title: Recognition, Translation, Coexistence*, CAEPR Research Monograph No. 27, ANU E Press, Canberrra.

Reeves, J. 1998. *Building on Land Rights for the Next Generation: The review of the Aboriginal Land Rights (Northern Territory) Act 1976*, AGPS, Canberra.

Render, J. 2005. *Mining and Indigenous Peoples Issues Review*, International Council on Mining and Metals, London.

Renzio, de Paolo. 2000. 'Bigmen and *Wantoks*: Social capital and group behaviour in Papua New Guinea', Working Paper Number 27, *QEH Working Paper Series,* Oxford Department of International Development, Oxford, viewed 11 December 2008, <http://www3.qeh.ox.ac.uk/pdf/qehwp/qehwps27.pd>.

Reserve Bank of Australia 2009. 'The level and distribution of recent mining sector revenue', *Reserve Bank Bulletin—January 2009,* viewed 14 April 2009, <http://www.rba.gov.au/PublicationsAndResearch/Bulletin/bu_jan09/the_lev_dis_mining_sec_rev.html>.

Rio Tinto Limited 2001. 'Aboriginal Policy and Programs—Briefing Note', Rio Tinto Limited, Melbourne.

Rio Tinto Iron Ore (RTIO) 2006. Pilbara Operations Sustainable Development Summary Report: More Value with Less Impact, RTIO, viewed 3 July 2008, <http://www.pilbarairon.com.au/sd/RTIO%20Sust%20Dev%20Summary%20(FA).pdf>.

—— 2007. 'New body provides representation for traditional owners', RTIO media release, 26 September, viewed 3 July 2008, <http://www.riotintoironore.com/ENG/media/38_media_releases_1378.asp>.

Roberts, J. 1978. *From Massacres to Mining: The Colonization of Aboriginal Australia*, CIMRA and War on Want, London.

Robinson, D. and Sidoti, C. 2000. 'The status of human rights in Australia', in S. Rees and S. Wright (eds), *Human Rights, Corporate Responsibility—A Dialogue*, Pluto Press, Annandale.

Rose, C. 2002. 'Corporate Responsibility, is it worth it and for whom?', Summary of a debate held in Brussels 9 December 2002, The Ethical Corporation, London, viewed 5 May 2008, <http://www.ethicalcorp.com>.

Ross, H. 1990. 'Progress and prospects in Indigenous social impact assessment', *Australian Indigenous Studies*, 1990/1: 11–17.

Ross, M. L. 1999. 'The political economy of the resource curse', *World Politics*, 51: 297–322.

Rowley, C. D. 1966. 'Discussion of papers by Messrs. de Vos, Gruen and Evans', in I. G. Sharp and C. M. Tatz (eds), *Aborigines in the Economy: Employment, Wages and Training*, Jacaranda Press, Brisbane.

—— 1971. *The Remote Aborigines*, ANU Press, Canberra.

—— 1972. *Outcasts in White Australia*, Penguin Books, Harmondsworth.

Rowse, T. 1992. *Remote Possibilities: the Aboriginal Domain and the Administrative Imagination*, North Australia Research Unit, ANU, Casuarina.

—— 1993. 'The principles of Aboriginal pragmatism', *Australian Quarterly*, 65 (4): 185–93.

—— 2000a. *Obliged to be Difficult: Nugget Coombs' Legacy in Indigenous Affairs*, Cambridge University Press, Cambridge.

—— 2000b. 'Culturally appropriate indigenous accountability', *American Behavioral Scientist*, 43 (9): 1514–32.

—— 2002. *Indigenous Futures: Choice and Development for Aboriginal and Islander Australia*, UNSW Press, Sydney.

—— 2005. 'The indigenous sector', in D. Austin-Broos and G. Macdonald (eds), *Culture, Economy and Governance in Aboriginal Australia*, Sydney University Press, Sydney.

—— 2006. 'The politics of being "practical": Howard's fourth term challenge', in T. Lea, E. Kowal and G. Cowlishaw (eds), *Moving Anthropology: Critical Indigenous Studies*, Charles Darwin University Press, Darwin.

—— 2008. 'The politics of "the gap" in Australia and New Zealand', Seminar presented 28 May, CAEPR, ANU, Canberra, Podcast <http://www.anu.edu.au/caepr/events08.php>.

Rumsey, A. and J. Weiner (eds) 2004. *Mining and Indigenous Lifeworlds in Australia and Papua New Guinea,* Sean Kingston Publishing, Wantage (UK).

Sahlins, M. 1999. 'What is anthropological enlightenment? Some lessons of the twentieth century', *Annual Review of Anthropology,* 28: i–xxiii.

Sanders, W. 1993. 'Rethinking the fundamentals of social policy towards indigenous Australians: Block grants, mainstreaming and the multiplicity of agencies and programs', *CAEPR Discussion Paper No. 46,* CAEPR, ANU, Canberra.

—— 2005. 'CDEP and ATSIC as bold experiments in governing differently: but where to now?', in D. Austin-Broos and G. Macdonald (eds), *Culture, Economy and Governance in Aboriginal Australia,* Sydney University Press, Sydney.

——, Taylor, J. and Ross, K. 2000. 'Participation and representation in ATSIC elections: A ten-year perspective', *CAEPR Discussion Paper No. 198,* CAEPR, ANU, Canberra.

Saward, M. 2003. 'Enacting democracy', *Political Studies,* 51 (1): 161–79.

Sawyer, S. and Gomez, E. T. 2008. 'Transnational governmentality and resource extraction: Indigenous peoples, multinational corporations, multilateral institutions and the state', *Identities, Conflict and Cohesion Programme Paper Number 13*, United National Research Institute for Social Development, Geneva.

Scambary, B. 2007. My Country, Mine Country: Indigenous People, Mining and Development Contestation in Remote Australia, PhD Thesis, ANU, Canberra.

Schwab, R. G. 1995. 'The calculus of reciprocity: Principles and implications of Indigenous sharing', *CAEPR Discussion Paper No. 100,* CAEPR, ANU, Canberra.

Scott, J. 1998. *Seeing Like a State: How Certain Schemes to Improve the Human Condition have Failed,* Yale University Press, New Haven.

Sen, A. 1999. *Development as Freedom,* Oxford University Press, London.

—— and Anand, S. 1994. Sustainable Human Development: Concepts and Priorities, Human Development Reports, viewed 14 March 2005, <http:/hdr.undp.org/docs/publications/ocational_papers/Oc8b.htm>.

Senior, C. 1998. 'The Yandicoogina process: A model for negotiating land use agreements', *Land, Rights, Laws: Issues of Native Title*, Regional Agreements Paper No. 6, Native Title Research Unit, AIATSIS, Canberra.

—— 2000. 'The Yandicoogina process: A model for negotiating Land Use Agreements', in P. Moore (ed), *Land, Rights, Laws Issues of Native Title*, Vol. 1, Native Title Research Unit, AIATSIS, Canberra.

Siegel, J. S. 2002. *Applied Demography: Applications to Business, Government, Law and Public Policy*, Academic Press, San Diego.

Siopsis, J., Hon. 2007. Gumala Investments Pty Ltd v Lethbridge [2007], Federal Court of Australia 934, 14 June 2007, viewed 11 December 2008, <http://www.austlii.edu.au/cgi-bin/sinodisp/au/cases/cth/FCA/2007/934.html?query=Gumala>.

Smith, D. E. 1998. 'Indigenous land use agreements: The opportunities, challenges and policy implications of the amended Native Title Act', *CAEPR Discussion Paper No. 163*, CAEPR, ANU, Canberra.

—— 2001. 'Valuing native title: Aboriginal, statutory and policy discourses about compensation', *CAEPR Discussion Paper No. 222*, CAEPR, ANU, Canberra.

—— 2002. 'Jurisdictional devolution: Towards an effective model for Indigenous community self-determination', *CAEPR Discussion Paper No. 233*, CAEPR, ANU, Canberra.

—— 2005. 'Indigenous families, households and governance', in D. Austin-Broos and G. Macdonald (eds), *Culture, Economy and Governance in Aboriginal Australia*, Sydney University Press, Sydney.

—— and Hunt, J. 2008. 'Understanding Indigenous Australian governance—research, theory and representations', in J. Hunt, D. E. Smith, S. Garling, and W. Sanders (eds), *Contested Governance: Culture, Power and Institutions in Indigenous Australia*, CAEPR Research Monograph No. 29, ANU E Press, Canberra.

Smith, L. T. 1999. *Decolonizing Methodologies: Research and Indigenous Peoples*, Zed Books, London and New York.

Smith, S. K. and Sincich, T. 1991. 'An empirical analysis of the effect of length of forecast horizon on population forecast errors', *Demography*, 28 (2): 261–74.

Solomon, F. 2000. 'Zen and the art of stakeholder involvement', *Occasional Paper No. 12*, Australian Minerals and Energy Environment Foundation, Melbourne.

Sosa, I. and Keenan, K. 2001. 'Impact benefit agreements between Aboriginal communities and mining companies: their use in Canada', Canadian Environmental Law Association, Environmental Mining Council of British Columbia, CooperAcción, viewed 11 December 2008, <http://cela.ca/uploads/f8e04c51a8e04041f6f7faa046b03a7c/IBAeng.pdf>.

Spar, D. 1998. 'The spotlight and the bottom line', *Foreign Affairs,* March/April, 77: 2.

Stanner, W. E. H. 1979 [1958]. 'Continuity and change (1958)', in *White Man Got No Dreaming, Essays 1938–1973*, ANU Press, Canberra.

Strehlow, T. G. H. n.d. *Dark and White Australians*, pamphlet.

Steering Committee for the Review of Government Service Provision 2003. *Overcoming Indigenous Disadvantage: Key Indicators 2003*, Productivity Commission, Canberra.

—— 2005. *Overcoming Indigenous Disadvantage: Key Indicators 2005*, Productivity Commission, Canberra.

Stokes, G. 2002. 'Australian democracy and Indigenous self-determination, 1901–2001', in G. Brennan and F. Castles (eds), *Australia Reshaped: 200 Years of Institutional Transformation,* Cambridge University Press, Cambridge.

Storey, A. 2003. 'Measuring development', in G. McCann and S. McCloskey (eds), *From the Local to the Global: Key Issues in Development Studies,* Pluto Press, London.

Storey, K. 2001. 'Fly-in/fly-out and fly-over: Mining and regional development in Western Australia', *Australian Geographer,* 32 (2): 133–48.

Strelein, L. 2003. 'Members of the Yorta Yorta Aboriginal Community v Victoria [2002] HCA 58 (12 December 2002)—Comment', in *Land, Rights, Laws: Issues of Native Title*, Vol. 2, Issues Paper 21, AIATSIS, Canberra.

—— 2006. *Compromised Jurisprudence: Native Title Cases Since Mabo*, Aboriginal Studies Press, Canberra.

Suggett, D. 2000. 'The rise of socially responsible investment', *Corporate Public Affairs,* 10 (2), viewed 17 December 2008, <http://www.accpa.com.au/login.php>.

Sullivan, P. 1996. *All Free Man Now: Culture, Community and Politics in the Kimberley Region, North-Western Australia*, AIATSIS Report Series, Canberra.

—— 1997. 'A sacred land, a sovereign people, an Aboriginal corporation: Prescribed bodies and the Native Title Act', *North Australia Research Unit Report Series No. 3*, North Australia Research Unit, ANU, Casuarina.

—— 2006. 'Indigenous governance: The Harvard Project on Native American Economic Development and appropriate principles of governance for Aboriginal Australia', *Research Discussion Paper No. 17*, AIATSIS, Canberra.

—— 2007. 'Indigenous governance: The Harvard Project, Australian Aboriginal organisations and cultural subsidiarity', *Working Paper No. 4*, Desert Knowledge Cooperative Research Centre, Alice Springs.

Sutton, P. 1995. *Country: Aboriginal Boundaries and Land Ownership in Australia*, Aboriginal History Monograph No. 3, Aboriginal History Inc., Canberra.

—— 1998. 'Families of polity: Post-classical Aboriginal society and native title', in P. Sutton, *Native Title and the Descent of Rights*, National Native Title Tribunal, Perth.

—— 2001. 'The politics of suffering: Aboriginal policy in Australia since the 1970s', *Anthropological Forum*, 11 (2): 125–73.

—— 2003. *Native Title in Australia: An Ethnographic Perspective*, Cambridge University Press, Cambridge.

Tatz, C., Cass, A., Condon, J. and Tipett, G. 2006. 'Aborigines and uranium: Monitoring the health hazards', *AIATSIS Research Discussion Paper No. 20*, AIATSIS, Canberra.

Taylor, J. 1993. 'Census enumeration in remote Australia: Issues for Aboriginal data analysis', *Journal of the Australian Population Association*, 10 (1): 53–67.

—— 1999. 'Aboriginal people in the Kakadu region: Social indicators for impact assessment', *CAEPR Working Paper No. 4*, CAEPR, ANU, Canberra.

—— 2003. 'Indigenous Australians: The first transformation', in S.E. Khoo and P. McDonald (eds), *The Transformation of Australia's Population: 1970-2030*, UNSW Press, Sydney.

—— 2004a. *Aboriginal Population Profiles for Development Planning in the Northern East Kimberley*, CAEPR Research Monograph No. 23, ANU E Press, Canberra.

—— 2004b. *Social Indicators for Aboriginal Governance: Insights from the Thamarrurr Region, Northern Territory*, CAEPR Research Monograph No. 24, ANU E Press, Canberra.

—— 2005. 'Indigenous labour supply and regional industry', in D. Austin-Broos, and G. Macdonald (eds), *Culture and Economy in Aboriginal Australia*, Sydney University Press, Sydney.

——— 2006a. 'Population and diversity: Policy implications or emerging Indigenous demographic trends', *CAEPR Discussion Paper No. 283*, CAEPR, ANU, Canberra.

——— 2006b. 'Indigenous people in the West Kimberley labour market', *CAEPR Working Paper No. 35*, CAEPR, ANU, Canberra.

——— 2008a. 'Indigenous demography and public policy: From overcoming Indigenous disadvantage to overcoming disadvantaged indicators', Plenary address to *Demographic Change in the 21st Century*, 14th Biennial Conference of the Australian Population Association, 1–2 July, Alice Springs.

——— 2008b. 'Indigenous labour supply constraints in the West Kimberley', *CAEPR Working Paper No. 39*, CAEPR, ANU, Canberra.

——— 2008c. 'Indigenous peoples and indicators of well-being: Australian perspectives on United Nations global frameworks', *Social Indicators Research*, 87: 111-26.

——— and Bell, M. 2001. Implementing Regional Agreements: Aboriginal Population Projections in Rio Tinto Mine Hinterlands 1996–2016, Report to Rio Tinto Ltd, Melbourne.

——— and ——— 2003. 'Options for benchmarking ABS population estimates for Queensland Aboriginal and Torres Strait Islander communities', *CAEPR Discussion Paper No. 243*, CAEPR, ANU, Canberra.

———, Brown, D. and Bell, M. 2006. *Population Dynamics and Demographic Accounting in Arid and Savanna Australia: Methods, Issues and Outcomes*, Desert Knowledge Cooperative Research Centre Research Report No. 16, Desert Knowledge Cooperative Research Centre, Alice Springs.

——— and Scambary, B. 2005. *Indigenous People and the Pilbara Mining Boom: A Baseline for Regional Participation*, CAEPR Research Monograph No. 25, ANU E Press, Canberra.

——— and Stanley, O. 2005. 'The opportunity costs of the status quo in the Thamarrurr Region', *CAEPR Working Paper No. 28*, CAEPR, ANU, Canberra.

Tedesco L., Fainstein, M. and Hogan, L. 2003. *Indigenous People in Mining*, Australian Bureau of Agricultural and Resource Economics eReport 03.19, Canberra, viewed 17 December 2008, <http://www.abare.gov.au/publications_html/research/research_03/er_indigenouspeople.pdf>.

Tehan, M., Palmer, L., Langton, M., and Mazel, O. 2006. 'Sharing land and resources: Modern agreements and treaties with indigenous peoples in settler states', in M. Langton, O. Mazel, L. Palmer, K. Shain, and M. Tehan (eds), *Settling With Indigenous People*, The Federation Press, Sydney.

The Wilderness Society and The Mineral Policy Institute 1998. 'Major investors flee Jabiluka project', Media release, The Wilderness Society and The Mineral Policy Institute, Melbourne.

Thorburn, K. 2006. 'The limits to accountability within Indigenous community-based organisations: Reflections on fieldwork in the West Kimberley', Seminar presented 4 October, CAEPR, ANU, Canberra, Podcast <http://www.anu.edu.au/caepr/events.php#sem2>.

Throsby, D. 2001. *Economics and Culture,* Cambridge University Press, Cambridge.

Tiplady, T. and Barclay, M. A. 2007. *Indigenous Employment in the Australian Minerals Industry*, Centre for Social Responsibility in Mining, University of Queensland, Brisbane.

Tonkinson, R. 1984–85. 'Councils, corporations and the Aboriginal polity', *Anthropological Forum*, 5 (3): 377–81.

—— 2007. 'Aboriginal "difference" and "autonomy" then and now: Four decades of change in a Western Desert society', *Anthropological Forum,* 17 (1): 41–60.

Toohey, J. 1984. *Seven Years On, Report by Mr Justice Toohey to the Minister for Aboriginal Affairs on the Aboriginal Land Rights (Northern Territory) Act 1976 and Related Matters*, AGPS, Canberra.

Toussaint, S., Sullivan, P., Yu, S. and Mularty, M. 2001. Fitzroy Valley Cultural Values Study: A Preliminary Assessment, A Report for the Water and Rivers Commission, Centre for Anthropological Research, The University of Western Australia, Nedlands.

Trebeck, K. 2003. 'Corporate social responsibility, Indigenous Australians and mining', *The Australian Chief Executive*, April: 46–7.

—— 2005. Democratisation through Corporate Social Responsibility? The Case of Miners and Indigenous Australians, PhD Thesis, ANU Canberra.

—— 2007a. 'Tools for the disempowered? Indigenous leverage over mining companies', *Australian Journal of Political Science*, 42 (4): 541–62.

—— 2007b. 'Private sector contribution to regeneration: Concepts, actions and synergies', *CPPR Working Paper 9*, Centre for Public Policy for Regions, University of Glasgow, Glasgow.

——2008a. 'Exploring the responsiveness of companies in practice: Corporate social responsibility to stakeholders', *Social Responsibility Journal*, 4 (3): 349–65.

—— (forthcoming) 2009. 'Indigenous/NGO alliances confronting corporate/state alliances: The case of the Jabiluka uranium prospect', in R. Raman (ed.), *CSR: Discourses, Practices and Perspectives*, Palgrave Macmillan, Basingstoke.

Trigger, D. 1997a. 'Mining, landscape and the culture of development ideology in Australia', *Ecumene*, 4 (2): 161–80.

—— 1997b. 'Reflections on Century Mine: Preliminary thoughts on the politics of Aboriginal responses', in D. E. Smith and J. Finlayson (eds), *Fighting Over Country: Anthropological Perspectives*, CAEPR Research Monograph No. 12, CAEPR, ANU, Canberra.

—— 1998. 'Citizenship and Aboriginal responses to mining in the Gulf country', in N. Peterson and W. Sanders (eds), *Citizenship and Aboriginal Australians: Changing Conceptions and Possibilities*, Cambridge University Press, Cambridge.

—— 2005. 'Mining projects in remote Australia: Sites for the articulation on contesting of economic and cultural futures', in D. Austin-Broos and G. Macdonald (eds), *Culture, Economy and Governance in Aboriginal Australia*, Sydney University Press, Sydney.

Triggs, G. 2002. 'The rights of indigenous peoples to participate in resource development: An international legal perspective', in D. Zillman, A. Lucas and G. Pring (eds), *Human Rights in Natural Resource Development: Public Participation in the Sustainable Development of Mining and Energy Resources*, Oxford University Press, Oxford.

Turnbull, S. 1979. *Economic Development of Aboriginal Communities in the Northern Territory: Second Report, Self-sufficiency (with Land Rights)*, AGPS, Canberra.

United Nations 2006. *Report of the Meeting on Indigenous Peoples and Indicators of Well-Being*, United Nations Economic and Social Council, E/C.19/2006/CRP.3, New York.

'Update: Yvonne Margarula convicted of trespassing on her own land' 1998, *Indigenous Law Bulletin*, No. 69, viewed 7 January 2008, <http://www.austlii.edu.au/au/journals/ILB/1998/69.html>.

Urteage-Crovetto, P. 2008. The Camisea Project: The Peruvian State, the Indigenous Peoples and the IDB, Peruvian case study for the 'Transnational Governmentality of Resource Extraction' project, United Nations Research Institute for Social Development.

Usher, P. J., Duhaime, G. and Searles, E. 2003. 'The household as an economic unit in Arctic Aboriginal communities, and its measurement by means of a comprehensive survey', *Social Indicators Research*, 61: 175–202.

Vachon, D. and Toyne, P. 1983. 'Mining and the challenge of land rights', in N. Peterson and M. Langton (eds), *Aborigines, Land and Land Rights*, Australian Institute of Aboriginal Studies, Canberra.

van de Bund, J. 1996. 'An analysis of a successful training and employment programme with a major mining company', Paper presented at the *Doing Business with Aboriginal Communities* Conference, 27–29 February, Darwin.

Vanstone, A. The Hon. 2005. National Press Club Address, 23 February, Canberra, viewed 13 March 2007, <http://www.atsia.gov.au/media/former_minister/speeches/2005/23_02_2005_pressclub.aspx>.

—— 2007. 'Beyond conspicuous compassion: Indigenous Australians deserve more than good intentions', in J. Wanna (ed.), *A Passion for Policy: Essays in Public Sector Reform*, ANU E Press, Canberra.

Vidler, P. 2007. *Indigenous Employment and Business Development in the Queensland Resources Sector*, Report to the Queensland Resources Council, Centre for Social Responsibility in Mining, University of Queensland, Brisbane.

von Sturmer, J. 1982. 'Aborigines in the uranium industry: toward self-management in the Alligator River region?', in R. M. Berndt (ed.), *Aboriginal Sites, Rights and Resource Development*, University of Western Australia Press, Nedlands.

—— 1984a. 'A critique of the Fox Report', in Australian Institute of Aboriginal Studies, *Aborigines and Uranium: Consolidated Report on the Social Impact of Uranium Mining on the Aborigines of the Northern Territory*, AGPS, Canberra.

—— 1984b. 'The social impact of mining: economic consequences: residence, resources, and inequities', in Australian Institute of Aboriginal Studies, *Aborigines and Uranium: Consolidated Report on the Social Impact of Uranium Mining on the Aborigines of the Northern Territory*, AGPS, Canberra.

Walsh, F. and Mitchell, P. 2002. *Planning for Country: Cross-Cultural Approaches to Decision-Making on Aboriginal Lands*, IAD Press, Alice Springs.

Warren, R. 1999. 'Company legitimacy in the new millennium', *Business Ethics: A European Review,* 8 (4): 214–24.

Webley, S. 2001. 'Business ethics: A SWOT exercise', *Business Ethics: A European Review,* 10 (3): 267–71.

Weiner, J. and Glaskin, K. 2006. 'The re-invention of Indigenous laws and customs', *The Asia Pacific Journal of Anthropology (Special Issue: Custom: Indigenous Tradition and Law in the Twenty-First Century)*, 7 (1): 1–14.

—— and —— (eds) 2007. *Customary Land Tenure and Registration in Indigenous Australia and Papua New Guinea: Anthropological Perspectives*, ANU E Press, Canberra.

Wereta, W. and Bishop, D. 2006. 'Towards a Maori statistics framework', in J. P. White, S. Wingert, D. Beavon, and P. Maxim (eds), *Aboriginal Policy Research: Moving Forward Making a Difference*, Thompson Educational Publishing, Toronto.

Westbury, N. 2003. 'The Moree Aboriginal Employment Strategy: Partnerships in delivering employment outcomes through changing community relations—Indigenous Australia: Understanding concepts of culture and community', *Occasional Publication 3*, Cranlana, Melbourne.

Western Australian Council of Social Services 2000. *Model of Social Sustainability*, Western Australian Council of Social Services, Perth.

White, A. 1999. 'Sustainability and the accountable corporation: Society's rising expectations of business', *Environment*, 41 (8): 30–43.

Whitehead, L. 2002. *Democratization Theory and Experience*, Oxford University Press, Oxford.

Williams, I. (Executive General Manager—Mining, Pasminco Ltd) 1999. 'Century Zinc Project', Paper presented at *The Spirit of the Snowy—Fifty Years On*, The Australian Academy of Technological Sciences and Engineering Symposium, November, Canberra, available at <http://www.atse.org.au>.

Wilson, J. 1961. Authority and Leadership in a 'New Style' Australian Aboriginal Community: Pindan, Western Australia, Masters thesis, University of Western Australia, Perth.

—— 1980. 'The Pilbara Aboriginal social movement: An outline of its background and significance', in R. M. Berndt and C. H. Berndt (eds), *Aborigines of the West: Their Past and Their Present*, University of Western Australia Press, Perth.

Wolfe, P. 1999. *Settler Colonialism and the Transformation of Anthropology: The Politics and Poetics of an Ethnographic Event*, Cassell, London and New York.

Woodward, A. E. 1974a. *Aboriginal Land Rights Commission: First Report, July 1973*, AGPS, Canberra.

—— 1974b. *Aboriginal Land Rights Commission: Second Report, April 1974*, AGPS, Canberra.

Woolcock, M. 1998. 'Social capital and economic development: Toward a theoretical synthesis and policy framework', *Theory and Society*, 27 (2): 151–208.

—— and Narayan, D. 2000. 'Social capital: Implications for development theory, research, and policy', *World Bank Research Observer*, 15 (2): 225–49.

World Economic Forum 2004. 'Values and value: Communicating the strategic importance of corporate citizenship to investors', Survey findings, World Economic Forum, Geneva.

Young, I. 2000. *Inclusion and Democracy*, Oxford University Press, New York.

Young, E. 1981. *The Aboriginal Component in the Australian Economy: Tribal Communities in Rural Areas*, Development Studies Centre, ANU, Canberra.

Young, L. (with Vitenbergs, A.) 2007. *Lola Young Medicine Woman and Teacher,* Fremantle Arts Centre Press, Fremantle.

Zadek, S. 2003. *The Civil Corporation,* Earthscan, London.

Key publications from the 'Indigenous community organisations and miners: Partnering sustainable regional development?' project (to July 2009)

Altman, J. C. 2002. 'Partnering sustainable regional development: Indigenous community organisations and miners', *Australian Chief Executive*, July: 46.

Altman, J. C. 2004. 'Economic development and Indigenous Australia: Contestations over property, institutions and ideology', *The Australian Journal of Agricultural and Resource Economics*, 18 (3): 513–34.

Altman, J. C. and Dillon, M. C. 2005. 'Commercial development and natural resource management on the Indigenous estate: A profit-related investment proposal', *Economic Papers*, 24 (3): 249–62.

Altman, J. C., Linkhorn, C. and Clarke, J., assisted by B. Fogarty and K. Napier. 2005. *Land Rights and Development Reform in Remote Australia*, Oxfam Australia, Melbourne.

Holcombe, S. 2004. 'Early Indigenous engagement with mining in the Pilbara: Lessons from a historical perspective', *CAEPR Working Paper No. 24*, CAEPR, ANU, Canberra.

Holcombe, S. 2005. 'Indigenous organisations and miners in the Pilbara, Western Australia: Lessons from a historical perspective', *Journal of Aboriginal History*, 29: 107–35.

Holcombe, S. 2006. '"Community Benefit Packages": Development's encounter with pluralism in the case of the mining industry', in T. Lea, E. Kowal and G. Cowlishaw (eds), *Moving Anthropology: Critical Indigenous Studies*, Charles Darwin University Press.

Levitus, R. 2005. 'Land rights and local economies: The Gagudju Association and the mirage of collective self-determination', in D. Austin-Broos and G. Macdonald (eds), *Culture, Economy and Governance in Aboriginal Australia*, University of Sydney Press, Sydney.

Levitus, R. 2007. 'Laws and strategies: The contest to protect Aboriginal interests at Coronation Hill', in J. F. Weiner and K. Glaskin (eds), *Customary Land Tenure and Registration in Australia and Papua New Guinea: Anthropological Pers*pectives, ANU E Press, Canberra.

Martin, D. F. 2005. 'Enhancing and measuring social sustainability by the minerals industry: A case study of Australian Aboriginal People',

Sustainable Development Indicators in the Minerals Industry, Proceedings of the Aachen International Mining Symposia, RWTH Aachen University, Institute of Mining Engineering, Aachen University, Germany.

Martin, D. F. 2005. 'Rethinking Aboriginal community governance: Challenges for sustainable engagement', in *Community and Local Governance in Australia*, P. Smyth, T. Reddel and A. Jones (eds), University of New South Wales Press, Sydney.

Martin, D. F. 2005. 'Governance, cultural appropriateness and accountability within the context of Indigenous self–determination', in D. Austin-Broos and G. Macdonald (eds), *Culture, Economy and Governance in Aboriginal Australia*, University of Sydney Press, Sydney.

Martin, D. F. and Finlayson, J. D. 2006. 'Regulating difference: Aborigines in the settler state', in P. Beilharz and T. Hogan (eds), *Introducing Sociology—Place, Time and Division*, Oxford University Press, Melbourne.

Scambary, B. 2007. '"No vacancies at the Starlight Motel": Larrakia identity and the claims process', in B. R. Smith and F. Morphy, *The Social Effects of Native Title: Recognition, Translation, Coexistence*, CAEPR Research Monograph No. 27, ANU E Press, Canberra.

Taylor, J. 2003. *Aboriginal Population Profiles for Development Planning in the Northern East Kimberley*, CAEPR Research Monograph No. 23, CAEPR, ANU, Canberra, xviii+124pp.

Taylor, J. 2003. 'Indigenous economic futures in the Northern Territory: The demographic and socioeconomic background', *CAEPR Discussion Paper No. 246*, CAEPR, ANU, Canberra.

Taylor, J. 2004. *Aboriginal Population Profiles for Development Planning in the Northern East Kimberley*, CAEPR Research Monograph No. 23, ANU E Press, ANU, Canberra.

Taylor, J. and Scambary, B. 2005. *Indigenous People and the Pilbara Mining Boom: A Baseline for Regional Participation*, CAEPR Research Monograph No. 25, ANU E Press, Canberra.

Trebeck, K. A. 2003. 'Corporate social responsibility, Indigenous Australians and mining', *Australian Chief Executive*, April: 46–7, 54.

Trebeck, K. A. 2004. 'Companies, complexity and CSR', *Australian Chief Executive*, April 2004: 48–9.

Trebeck, K. A. 2004. 'Democracy and community instigated CSR', *The Corporate Citizen,* 4 (1): 11–13.

Trebeck, K. A. 2007. 'Tools for the disempowered? Indigenous leverage over mining companies', *Australian Journal of Political Science*, 42 (4): 541–62.

Trebeck, K. A. 2008. 'Corporate social responsibility and democratisation: Opportunities and obstacles', in C. O'Faircheallaigh and S. Ali (eds), *Indigenous Peoples, the Extractive Industries and Corporate Social Responsibility*, Greenleaf Publishing, Sheffield.

Trebeck, K. A. 2008. 'Exploring the responsiveness of companies: corporate social responsibility to stakeholders', *Social Responsibility Journal*, 4 (3): 349–65.

Trebeck, K. A. 2008. 'Relative advantages: Exploring private sector impact on disadvantaged groups and deprived areas', *Journal of Corporate Citizenship,* 32: 1–17.

Trebeck, K. A. forthcoming. *'Indigenous/NGO alliances confronting corporate/state alliances: The case of the Jabiluka uranium prospect'*, in R. Raman (ed.), *CSR: Discourses, Practices and Perspectives*, Palgrave Macmillan, Basingstoke.

CAEPR Research Monograph Series

1. *Aborigines in the Economy: A Select Annotated Bibliography of Policy Relevant Research 1985–90*, L. M. Allen, J. C. Altman, and E. Owen (with assistance from W. S. Arthur), 1991.

2. *Aboriginal Employment Equity by the Year 2000*, J. C. Altman (ed.), published for the Academy of Social Sciences in Australia, 1991.

3. *A National Survey of Indigenous Australians: Options and Implications*, J. C. Altman (ed.), 1992.

4. *Indigenous Australians in the Economy: Abstracts of Research, 1991–92*, L. M. Roach and K. A. Probst, 1993.

5. *The Relative Economic Status of Indigenous Australians, 1986–91*, J. Taylor, 1993.

6. *Regional Change in the Economic Status of Indigenous Australians, 1986–91*, J. Taylor, 1993.

7. *Mabo and Native Title: Origins and Institutional Implications*, W. Sanders (ed.), 1994.

8. *The Housing Need of Indigenous Australians, 1991*, R. Jones, 1994.

9. *Indigenous Australians in the Economy: Abstracts of Research, 1993–94*, L. M. Roach and H. J. Bek, 1995.

10. *The Native Title Era: Emerging Issues for Research, Policy, and Practice*, J. Finlayson and D. E. Smith (eds), 1995.

11. *The 1994 National Aboriginal and Torres Strait Islander Survey: Findings and Future Prospects*, J. C. Altman and J. Taylor (eds), 1996.

12. *Fighting Over Country: Anthropological Perspectives*, D. E. Smith and J. Finlayson (eds), 1997.

13. *Connections in Native Title: Genealogies, Kinship, and Groups*, J. D. Finlayson, B. Rigsby, and H. J. Bek (eds), 1999.

14. *Land Rights at Risk? Evaluations of the Reeves Report*, J. C. Altman, F. Morphy, and T. Rowse (eds), 1999.

15. *Unemployment Payments, the Activity Test, and Indigenous Australians: Understanding Breach Rates*, W. Sanders, 1999.

16. *Why Only One in Three? The Complex Reasons for Low Indigenous School Retention*, R. G. Schwab, 1999.

17. *Indigenous Families and the Welfare System: Two Community Case Studies*, D. E. Smith (ed.), 2000.

18. *Ngukurr at the Millennium: A Baseline Profile for Social Impact Planning in South-East Arnhem Land*, J. Taylor, J. Bern, and K. A. Senior, 2000.

19. *Aboriginal Nutrition and the Nyirranggulung Health Strategy in Jawoyn Country*, J. Taylor and N. Westbury, 2000.

20. *The Indigenous Welfare Economy and the CDEP Scheme*, F. Morphy and W. Sanders (eds), 2001.

21. *Health Expenditure, Income and Health Status among Indigenous and Other Australians*, M. C. Gray, B. H. Hunter, and J. Taylor, 2002.

22. *Making Sense of the Census:Observations of the 2001 Enumeration in Remote Aboriginal Australia*, D. F. Martin, F. Morphy, W. G. Sanders and J. Taylor, 2002.

23. *Aboriginal Population Profiles for Development Planning in the Northern East Kimberley*, J. Taylor, 2003.

24. *Social Indicators for Aboriginal Governance: Insights from the Thamarrurr Region, Northern Territory*, J. Taylor, 2004.

25. *Indigenous People and the Pilbara Mining Boom: A Baseline for Regional Participation*, J. Taylor and B. Scambary, 2005.

26. *Assessing the Evidence on Indigenous Socioeconomic Outcomes: A Focus on the 2002 NATSISS*, B. H. Hunter (ed.), 2006.

27. *The Social Effects of Native Title: Recognition, Translation, Coexistence*, B. R. Smith and F. Morphy (eds), 2007.

28. *Agency, Contingency and Census Process: Observations of the 2006 Indigenous Enumeration Strategy in remote Aboriginal Australia*, F. Morphy (ed.), 2008.

29. *Contested Governance: Culture, Power and Institutions in Indigenous Australia*, Janet Hunt, Diane Smith, Stephanie Garling and Will Sanders (eds), 2008.

For information on CAEPR Discussion Papers, Working Papers and Research Monographs (Nos 1-19) please contact:

> Publication Sales, Centre for Aboriginal Economic Policy Research,
> College of Arts and Social Sciences,
> The Australian National University, Canberra, ACT, 0200

> Telephone: 02–6125 8211
> Facsimile: 02–6125 2789

Information on CAEPR abstracts and summaries of all CAEPR print publications and those published electronically can be found at the following WWW address: http://www.anu.edu.au/caepr/

www.ingramcontent.com/pod-product-compliance
Lightning Source LLC
Chambersburg PA
CBHW061244270326
41928CB00041B/3405